Sustainable Environmental Management

Other titles from the Society of Environmental Toxicology and Chemistry (SETAC):

Principles and Processes for Evaluating Endocrine Disruption in Wildlife
R.J. Kendall, R.L. Dickerson, J.P. Giesy, W.P. Suk, editors

Chemical Ranking and Scoring:
Guidelines for Relative Assessments of Chemicals
M.B. Swanson, A.B. Socha, editors

Chemically Induced Alterations in Functional Development and Reproduction of Fishes
R.M. Rolland, M. Gilbertson, R.E. Peterson, editors

Atmospheric Deposition of Contaminants to the Great Lakes and Coastal Waters
J.E. Baker, editor

Ecological Risk Assessment of Contaminated Sediments
C.G. Ingersoll, T. Dillon, G.R. Biddinger, editors

Reassessment of Metals Criteria for Aquatic Life Protection:
Priorities for Research and Implementation
H.L. Bergman and E.J. Dorward-King, editors

Whole Effluent Toxicity Testing:
An Evaluation of Methods and Prediction of Receiving System Impacts
D.R. Grothe, K.L. Dickson, D.K. Reed-Judkins, editors

The Multi-Media Fate Model:
A Vital Tool for Predicting the Fate of Chemicals
C.E. Cowan, D. Mackay, T.C.J. Feijtel, D. Van de Meent, A. DiGuardo,
J. Davies, N. Mackay, editors

Aquatic Dialogue Group: Pesticide Risk Assessment and Mitigation
J.L. Baker, A.C. Barefoot, L.E. Beasley, L. Burns, P. Caulkins, J. Clark,
R.L. Feulner, J.P. Giesy, R.L. Graney, R. Griggs, H. Jacoby, D. Laskowski,
A. Maciorowski, E. Mihaich, H. Nelson, R. Parrish, R.E. Siefert, K.R. Solomon,
W. van der Schalie, editors

For information about any SETAC publication, including SETAC's international journal
Environmental Toxicology and Chemistry,
contact the SETAC Office, 1010 N. 12th Avenue, Pensacola, Florida, USA 32501-3370
T 850 469 1500   F 850 469 9778   E setac@setac.org   http://www.setac.org

Sustainable Environmental Management

Edited by

Lawrence W. Barnthouse
LWB Environmental Services, Inc.

Gregory R. Biddinger
Exxon Company USA

William E. Cooper
Michigan State University

James A. Fava
Roy F. Weston Inc.

James H. Gillett
Cornell University

Marjorie M. Holland
University of Mississippi

Terry F. Yosie
Ruder Finn Company

Proceedings of the Pellston Workshop on Sustainability-Based Environmental Management
25–31 August 1993
Pellston, Michigan

SETAC Special Publications Series

SETAC Liaison
Greg Schiefer
Society of Environmental Toxicology and Chemistry (SETAC)

Current Coordinating Editor of SETAC Books
C. G. Ingersoll
U.S. Geological Survey

Publication sponsored by the Society of Environmental Toxicology
and Chemistry (SETAC) and the SETAC Foundation for Environmental Education

Cover by Michael Kenney Graphic Design and Advertising
Editing and typesetting by Wordsmiths Unlimited

Library of Congress Cataloging-in-Publication Data

Pellston Workshop on Sustainability-Based Environmental Management
 (1993 : Pellston, Mich.)
  Sustainable environmental management : proceedings of the Pellston Workshop on Sustainability-Based Environmental Management, 25–31 August, 1993. Pellston, Michigan / edited by Lawrence W. Barnthouse ... [et al.] : sponsored by U.S. Environmental Protection Agency ... [et al.].
   p. cm. -- (SETAC special publications series)
   ISBN 1-880611-08-2
   1. Environmental management--Congresses. I. Barnthouse, L. W. (Lawrence W.) II. Title. III. Series.
GE300.P45 1993
363.7--dc21   97-36153
                                                                                                                CIP

Information in this book was obtained from individual experts and highly regarded sources. It is the publisher's intent to print accurate and reliable information, and numerous references are cited; however, the authors, editors, and publisher cannot be responsible for the validity of all information presented here or for the consequences of its use. Information contained herein does not necessarily reflect the policy or views of the Society of Environmental Toxicology and Chemistry (SETAC).

No part of this publication may be reproduced, stored in a retrieval system, or transmitted in any form or by any means, electronic, electrostatic, magnetic tape, mechanical, photocopying, recording, or otherwise, without permission in writing from the copyright holder.

All rights reserved. Authorization to photocopy items for internal or personal use, or the personal or internal use of specific clients, may be granted by the Society of Environmental Toxicology and Chemistry (SETAC), provided that $9.00 per chapter photocopied is paid directly to Copyright Clearance Center, 222 Rosewood Drive, Danvers, MA 01923 USA (telephone 978-750-8400).

SETAC's consent does not extend to copying for general distribution, for promotion, for creating new works, or for resale. Specific permission must be obtained in writing from SETAC for such copying. Direct inquiries to the Society of Environmental Toxicology and Chemistry (SETAC), 1010 North 12th Avenue, Pensacola, FL 32501-3370, USA.

© 1998 Society of Environmental Toxicology and Chemistry (SETAC)
SETAC Press is an imprint of the Society of Environmental Toxicology and Chemistry.
No claim is made to original U.S. Government works.

International Standard Book Number 1-880611-08-2
Printed in the United States of America
05 04 03 02 01 00 99 98   10 9 8 7 6 5 4 3 2 1

∞ The paper used in this publication meets the minimum requirements of the American National Standard for Information Sciences—Permanence of Paper for Printed Library Materials, ANSI Z39.48-1984.

Reference Listing: Barnthouse LW, Biddinger GR, Cooper WE, Fava JA, Gillett JH, Holland MM, Yosie TF, editors. 1998. Sustainable environmental management. SETAC Pellston Workshop on Sustainability-Based Environmental Management; 1993 Aug 25–31; Pellston MI. Published by the Society of Environmental Toxicology and Chemistry (SETAC), Pensacola, Florida, USA. 128 p.

The SETAC Special Publications Series

The SETAC Special Publications Series was established by the Society of Environmental Toxicology and Chemistry (SETAC) to provide in-depth reviews and critical appraisals on scientific subjects relevant to understanding the impacts of chemicals and technology on the environment. The series consists of single- and multiple-authored or edited books on topics reviewed and recommended by the SETAC Board of Directors and approved by the Publications Advisory Council for their importance, timeliness, and contribution to multidisciplinary approaches to solving environmental problems. The diversity and breadth of subjects covered in the series reflect the wide range of disciplines encompassed by environmental toxicology, environmental chemistry, and hazard and risk assessment. Despite this diversity, the goals of these volumes are similar; they are to present the reader with authoritative coverage of the literature, as well as paradigms, methodologies and controversies, research needs, and new developments specific to the featured topics. All books in the series are peer reviewed for SETAC by acknowledged experts.

The SETAC Special Publications are useful to environmental scientists in research, research management, chemical manufacturing, regulation, and education, as well as to students considering careers in these areas. The series provides information for keeping abreast of recent developments in familiar subject areas and for rapid introduction to principles and approaches in new subject areas.

Sustainable Environmental Management presents the collected papers stemming from a SETAC- and Ecological Society of America (ESA)-sponsored Pellston Workshop on Sustainability-Based Environmental Management, held 25–31 August 1993 in Pellston, Michigan. The workshop focused on discussions of unresolved scientific issues and needed research in the area of sustainable environmental management (SEM). Like all previous SETAC workshops, participation was limited to invited experts from government, academia, and industry who were selected because of their experience with the workshop topic. The workshop provided a structured environment for the exchange of ideas and debate such that consensus positions would be derived and documented for some of the issues surrounding the science of SEM. Participants prepared chapters that not only presented new information but also reviewed and synthesized the current state of the science. Each chapter was peer reviewed by leading scientists and revised before the compiled book was extensively scrutinized for technical and editorial accuracy and consistency by the editors and by paid consultants.

SETAC would like to recognize the past SETAC Special Publications Series editors:
    T. W. La Point, The Institute for Environmental and Human Health,
        Texas Tech University, Lubbock TX
    B. T. Walton, U.S. Environmental Protection Agency,
        Research Triangle Park NC
    C. H. Ward, Department of Environmental Sciences and Engineering,
        Rice University, Houston TX

Contents

List of Figures vii
List of Tables vii
Foreword viii
Preface xii
Acknowledgments xiii
About the Editors xiv
Workshop Steering Committee, Plenary Speakers, and Discussion Groups xviii
Executive Summary xix
Abbreviations xxiii

Chapter 1: Introduction 1

Chapter 2: Issues in Sustainable Environmental Management 7
    Transboundary problems 7
    Landscape and watershed management 9
    Product stewardship 11
    Biodiversity protection and restoration 13
    Urban environmental management 16
    Environmental restoration 18

Chapter 3: Case Studies 19
    Integrated pest management (Joseph Kovach) 19
    Scientific and management issues for the Great Lakes (John Hartig) .. 22
    Sustainability-based forest management in Michigan's Upper
        Peninsula (Dean Premo) 28
    Product development (Sven-Olof Ryding) 33
    The Everglades (Larry Harris) 37

Chapter 4: Steps toward Sustainable Environmental Management 41
    Dissemination of sustainable technologies 41
    Strategy for sustainable management of regions 43
    Product stewardship 47
    Social choice mechanisms and biodiversity 57
    Ensuring sustainability of urban environments 62
    New approaches to environmental restoration 67

Chapter 5: Conclusions and Recommendations 81

Appendix A: Workshop Agenda 85
Appendix B: Workshop Participants and Contributing Authors 89

References 91
Index 95

List of Figures

Figure 1	Principal components of sustainable environmental management	2
Figure 2	Development of sustainable environmental management strategies	5
Figure 3	Strategy for sustainable management of regional resources	44
Figure 4	Framework for implementing sustainable product development	47
Figure 5	Typical energy and material flows for a product life cycle	48
Figure 6	Integrating sustainable environmental management philosophy into organizational operations	53
Figure 7	Implementing sustainable development strategies and tools	57
Figure 8	Ensuring sustainable development in urban environments	63
Figure 9	Development of a remedial action plan	65
Figure 10	Incorporating sustainability principles into site restoration planning	71
Figure 11	Adaptive management strategy	73
Figure 12	Adaptive management process for flexible restoration planning	74

List of Tables

Table 1	Social welfare impact categories	3
Table 2	Current integration of health and environmental issues within typical manufacturing companies	11
Table 3	Theoretical environmental impact of different pest management strategies used for Red Delicious apples in New York	21
Table 4	Sustainable development strategic matrix	55
Table 5	Comparison of conventional management and sustainable environmental management methodologies	68
Table 6	Potential stakeholders in adaptive restoration management	74

Foreword

Environmental protection and management have been major international priorities for only about 25 years. The environmental problems that must be addressed range from resource management concerns (e.g., fisheries management) to localized environmental contamination (water pollution and pesticide use) to regional degradation (acid deposition) and global change. These diverse problems historically have been addressed by independent groups of scientists. Management objectives have often conflicted. Scientists involved in basic ecological research and those involved in practical environmental management traditionally have belonged to different professional societies, whose agendas have little in common.

From a strictly technical perspective, during the last decade it has become increasingly clear that the major environmental problems of the 1990s, such as regional environmental contamination, habitat destruction, and climate change, require interdisciplinary cooperation and new approaches to environmental risk assessment and management. Observed cases of large-scale environmental deterioration usually involve multiple human influences that cannot be addressed piecemeal. For these reasons, it is attractive to manage ecological systems based on measurement of their status or quality and to define management goals in terms of maintaining system integrity (i.e., "sustainability"). Such approaches require types of information that have been used infrequently in ecological assessment and management.

Recently the demands placed on ecologists have been greatly transformed and intensified because of the worldwide movement by both industrialized and developing nations toward balancing environmental quality, economic growth, and cultural development. Successfully achieving environmentally sustainable development will require advances in basic understanding of the interaction between human activities and ecological systems and innovative approaches to using ecological knowledge in environmental management. Scientists will have to step outside their traditional roles as academic researchers, resource managers, and regulatory specialists to participate much more actively in translating the universally accepted but operationally vague objective of "sustainable development" into concrete (and achievable) management plans.

The need for fundamental advances in the science underlying ecological resource management has been recognized by two major environmental societies, the Society of Environmental Toxicology and Chemistry (SETAC) and the Ecological Society of America (ESA). SETAC sponsored a Pellston Workshop in Breckenridge, Colorado in 1987 on "Consensus Research Needs and Priorities in Environmental Risk Assessment" (Fava et al. 1987). The report identified broad areas of uncertainty in risk assessment methodology. General recommendations about the need for basic research were made, but specific recommendations emphasized aquatic toxicology, terrestrial toxicology, and chemistry, fate, and modeling. Subsequent workshops in 1989 and 1990 developed more detailed assessments of the state-of-the-science in biomarkers (Huggett et al. 1992) and avian population ecology and toxicology (Kendall and Lacher Jr 1993).

In 1990, ESA proposed a "Sustainable Biosphere Initiative" (SBI), identifying basic research priorities related to global change, biological diversity, and sustainable ecological systems. The sustainable systems component, which deals with management and restoration of ecosystems affected by diverse anthropogenic influences, largely overlaps the scope of environmental risk assessment outlined in the SETAC "Research Needs" report. General recommendations were made, but no specific research plan was proposed.

This report presents results from a jointly sponsored workshop intended to identify some general principles for bringing science to bear on sustainability problems, recognizing that the ways in which science and scientists actually influence environmental management are much more complex than many have previously understood.

Society of Environmental Toxicology and Chemistry

SETAC is a professional society of more than 5000 members, founded in 1979 to provide a forum for individuals and institutions engaged in the study of environmental problems, management and regulation of natural resources, education, research and development, and manufacturing and distribution. It is the only professional society that specifically brings together environmental scientists and engineers from academia, government, industry, and public interest groups to provide research, education, and training in environmental problem solving. SETAC provides a forum through meetings, publications, and workshops for communication among professionals involved in the use, protection, and management of our environment. The goals of SETAC are pursued through such activities as these:
- holding an annual meeting consisting of workshops and paper and poster presentations on topics related to environmental toxicology and chemistry;
- publishing a monthly journal *Environmental Toxicology and Chemistry*, a bimonthly newsletter, *SETAC News*, and special publications;
- organizing and sponsoring chapters to provide a forum for the presentation of scientific data and for the interchange and study of information of local concern; and
- providing advice and counsel to technical and nontechnical people, groups, and institutions about scientific issues through a number of standing ad hoc committees.

The Ecological Society of America

The Ecological Society of America (ESA) is a nonpartisan, nonprofit organization of scientists founded in 1915 for the purposes of unifying the science of ecology, stimulating research, encouraging communication among ecologists, and promoting the responsible application of ecological data and principles to the solution of environmental problems. The Society's nearly 7000 members in the United States, Canada, Mexico, and 62 other nations conduct research, teach, and work in government agencies and many other organizations. The Society publishes four journals: *Ecology, Ecological Monographs, Ecological Applications*, and *The Bulletin of the Ecological Society of America*. Annually the Society convenes a conference for the exchange of information; recent themes include ecological education and global sustainability.

The Society's Public Affairs Office, located in Washington DC, allows the Society to engage effectively in discussion of ecological issues and policies. The Public Affairs Office fulfills several functions:
- interacts regularly with federal agencies, Congressional offices, other scientific societies, environmental organizations, and the news media;
- serves as a liaison between ecologists and decision-makers; and
- provides access to scientific information on environmental and ecological issues of regional, national, and international importance.

Since the opening of the Office in 1983, the ESA has addressed a number of scientific issues, including loss of biological diversity, ecological consequences of introducing genetically engineered organisms, sustainability of Earth's resources, and wetland delineation. Upon request, ESA members have testified before Congressional committees on current scientific issues including biotechnology, global change, and maintenance of biological diversity.

The SBI Project Office, the second Washington-based ESA activity, is a product of efforts by the ecological science community to set research priorities for the next decade. A three-year consensus-building process among ESA members produced the "Sustainable Biosphere Initiative: An Ecological Research Agenda," which was published in the April 1991 issue of *Ecology* (Lubchenco et al. 1991).

The SBI document elaborates a research agenda for three priority areas: global change, biological diversity, and sustainability of ecological systems. The SBI grounds ecological research in the context of changing conditions: the increasing scarcity of research resources, the complex and interdisciplinary nature of most pressing environmental issues, and an urgent need for sound scientific input into the environmental policy-making process. Through the SBI, the ESA has provided a framework for the acquisition, dissemination, and utilization of ecological knowledge.

In early 1992, the National Science Foundation, in concert with other federal agencies (U.S. Forest Service, National Aeronautic and Space Administration [NOAA], U.S. Environmental Protection Agency [USEPA], U.S. Fish and Wildlife Service), funded the SBI Project Office, which offers unique tools for addressing issues of great scientific import and policy relevance. The SBI Project Office fills several roles:
- facilitates interaction between the scientific community and government agencies at all levels, Congress, national and international programs, nongovernmental organizations, industry, and other interested institutions and individuals;
- promotes investigator-initiated, peer-reviewed research and synthesis; promotes new research directions; assists in the development of research proposals

responding to agency programs and funding initiatives; facilitates scientific advisory and review processes;
- sponsors workshops, symposia, roundtables, action plans, and papers used by agencies to better understand ecological issues and develop programs and funding initiatives to address them; and
- develops products needed by decision-makers and managers in formulating and implementing environmental policy.

Through its activities, the SBI Project Office provides a powerful mechanism for creating innovative cooperative partnerships between the scientific community, other disciplines, and other constituencies actively searching for solutions to issues of global change, biological diversity, and sustainable ecological systems.

A Workshop on Sustainable Environmental Management

ESA and SETAC leaders realized that the difficult technical and institutional challenges would require the efforts of many groups working together. Thus, a plan was set in place to jointly sponsor a workshop. A steering committee that included members of both societies put together an agenda and identified speakers and invitees.

The Pellston Workshop on Sustainability-Based Environmental Management was held 25–31 August 1993 at the University of Michigan Biological Station in Pellston, Michigan. It was attended by 29 people, including 10 SETAC members, 4 ESA members, and 4 individuals who were members of both societies. Significantly, the workshop was also attended by 11 individuals who were members of neither society. These included an economist, several scientists and policy specialists, and a state environmental regulator. The steering committee, plenary speakers, and discussion groups are listed on page xviii. The workshop program and participant list are provided in Appendices A and B, respectively.

This report presents the results of the workshop. No definitive solutions to the problem of sustainability were provided, but significant issues of both a technical nature and a sociodynamic nature were raised and discussed. Perhaps surprisingly, much of the discussion focused not on technical issues regarding sustainable environmental management (SEM) but on the institutions and processes needed to link the scientific aspects of environmental management to the public decision-making process. The hope is to lay a foundation for continued cooperation not only between SETAC and ESA but among the scientific community, the business community, public interest groups, and environmental decision-makers at federal, state, and local levels.

Preface

This book presents the proceedings of the 26th Pellston Workshop, held 25–31 August 1993 in Pellston, Michigan. As in previous workshops, participation was limited to invited experts from government, academia, and industry, who were selected because of their experience with the workshop topic.

The Workshop on Sustainability-Based Environmental Management provided a structured environment for the exchange of ideas and debate such that consensus positions would be derived and documented for some of the issues surrounding the science and regulatory practice related to sustainable environmental management (SEM). The proceedings reflect the current state-of-the-art of these topics and focus on an assessment of 1) transboundary problems, 2) landscape and watershed management, 3) product stewardship, 4) biodiversity protection and restoration, 5) urban environmental management, and 6) environmental restoration.

Acknowledgments

The Workshop on Sustainability-Based Environmental Management and publication of the workshop proceedings were made possible through the financial support of the U.S. Environmental Protection Agency, the Chemical Manufacturers Association, Rohm and Haas Company, the Canadian Standards Association, Kennicott Corporation, and the Council for LAB/LAS Environmental Research. The content of this publication does not necessarily reflect the position or the policy of any of these organizations or the U.S. government, and no official endorsement should be inferred.

About the Editors

Larry Barnthouse is president and principal scientist of LWB Environmental Services, Inc. He was formerly a senior research staff member in Oak Ridge National Laboratory's Environmental Sciences Division, where for 19 years, he was involved in dozens of environmental research and assessment projects, including development of new methods for predicting and measuring environmental risks of energy technologies. After leaving Oak Ridge National Laboratory in 1995, he spent two and a half years with McLaren-Hart, Inc. Dr. Barnthouse has authored or coauthored more than 70 publications relating to ecological risk assessment. He is an internationally recognized authority on applications of population modeling techniques to problems involving power-plant cooling systems, toxic chemicals, and watershed management. Dr. Barnthouse has been a SETAC member since 1985. In addition to chairing the Workshop on Sustainability-Based Environmental Management, he has participated in Pellston workshops on product life-cycle assessment, chemical scoring, and reproductive effects of chemicals. He is the hazard/risk assessment editor of *Environmental Toxicology and Chemistry*.

Gregory R. Biddinger is presently an advisor for Exxon Company, USA in their Environmental and Safety Department. He obtained his doctoral training in Aquatic Ecology and Physiology at Indiana State University and subsequently trained as an environmental toxicologist while a post-doctoral associate at Cornell University. Since 1983, Dr. Biddinger has worked as an environmental toxicologist for the Illinois EPA and Exxon developing and directing programs to manage the risks of chemicals in the marketplace and the safe disposal of wastes from manufacturing processes. At the time of this workshop, Dr. Biddinger was the section head for ecotoxicology at Exxon Biomedical Science in East Millstone, New Jersey. During his career, Dr. Biddinger has been actively involved in the advancement and standardization of testing methods in environmental toxicology and risk assessment. Since 1992, he has chaired the SETAC Ecological Risk Assessment Advisory Group, which helped organize this workshop.

Bill Cooper is a full professor of zoology at Michigan State University, where he was department chairperson from 1981–1987. He became co-director of the Design and Management of Environmental Systems Project, sponsored by the Research Associated with National Needs (RANN) section of the National Science Foundation (NSF) in 1970, chaired the Michigan Environmental Review Board from 1975–1988, and is presently senior consultant for environmental science for Public Sector Consultants, Inc. He has been a member of the Science Advisory Board of the Great Lakes Center, (Michigan) Environmental Cabinet, Environmental Studies Board, National Academy of Sciences, (Michigan) Environmental Quality Council, and a consultant for the Michigan Aeronautics Commission; he is a member of 5 professional societies and 2 editorial boards and lectures at the Brookings Institute, Washington DC. He has had 2 terms on the National Research Council Board on Environmental Studies and Toxicology and is also the recipient of the Braun InterTec/Dow Chair in Civil Engineering at the University of Minnesota. Dr. Cooper is presently a driving force in the formation of the Hazardous Waste Management Consortium directed through the Institute for Environmental Toxicology at Michigan State University. He serves on the USEPA Science Advisory Board (SAB), has directed the Relative Risk Assessment Program for Michigan, and recently chaired the Michigan Environmental and Natural Resource Code Commission.

James Fava has 20 years' experience in environmental management, product stewardship, and regulatory policy activity. Dr. Fava is vice president for Strategic Management and Product Stewardship Services at Roy F. Weston, Inc., assisting clients to develop management systems, programs, and tools to include economic, environment, and social considerations into their decision-making processes. He currently chairs the ISO 14000 TC-207 U.S. Technical Advisory Group on Life-Cycle Assessment (LCA), contributes to the development of other ISO 14000 standards, and has led Product Stewardship Awareness sessions for numerous Fortune 500 companies and federal agencies. He has served as president of SETAC, is currently a member of the board of directors for the SETAC Foundation for Environmental Education, and chairs SETAC's LCA Advisory Group. Dr. Fava has delivered numerous technical papers and presentations on risk assessment, ISO, EMS implementation, strategic environmental management, product stewardship, LCA, DFE, and sustainable development to national and international audiences. He holds a Ph.D. in behavioral toxicology, a master of science degree in fisheries biology, and a bachelor of science degree in zoology, all from the University of Maryland in College Park, Maryland.

James W. Gillett is professor of ecotoxicology in the Department of Natural Resources at Cornell University and director of the Superfund Basic Research and Education Program. He has a B.S. degree in chemistry from the University of Kansas (1955) and a Ph.D. in biochemistry (1962) form the University of California at Berkeley, where he began his career in environmental toxicology of pesticides and toxic substances. He was in the Department of Agricultural Chemistry at Oregon State University before joining the USEPA where he was a senior scientist in the Office of Research and Development in the Corvallis Environmental Research Lab (1973–1983). Dr.' Gillett's research on physical and mathematical models of fate and transport have been directed toward both health and ecological risk assessments. Among the projects are municipal solid waste composting, work on subsistence exposure with the St. Regis Mohawk Indians, fish contamination by PCBs and DDT-R, and dermal and inhalation exposure from showering.

Marjorie M. Holland received an interdisciplinary Ph.D. through the Five-College Ph.D. Program, coordinated through the Botany Department at the University of Massachusetts in Amherst. Her dissertation work synthesized information and resources from the disciplines of fluvial geomorphology, plant systematics, riverine ecology, plant ecology, environmental history, and natural resource management. Dr. Holland has over a decade of experience in natural resource management and public policy, working on the local, state, regional, national, and international levels. In June 1987, she utilized her background in wetland science to serve as the Science Representative from the United Nations Educational, Scientific, and Cultural Organization (UNESCO) to the Third Meeting of the Conference of the Contracting Parties to the Convention on Wetlands of International Importance Especially as Waterfowl Habitat in Regina, Saskatchewan, Canada. Currently, she holds positions at the University of Mississippi as director of the university's Field Station and the Center for Water and Wetland Resources, and she is associate professor in the Biology Department on the Oxford, Mississippi campus.

Terry F. Yosie has spent nearly 2 decades at senior levels of U.S. government, private industry, and management consulting. Under his direction from 1981–1988, USEPA's Science Advisory Board (SAB) became one of the government's 10 largest advisory bodies. From 1988–1992, Dr. Yosie was a vice president at the American Petroleum Institute, where he managed health and environmental research programs, conducted legislative and regulatory analyses, provided strategic counseling on environmental management and policy, and served as principal strategist and spokesperson on public health and the environment. Since 1992, Dr. Yosie has served as executive vice president of E. Bruce Harrison Co., continuing in that capacity since the 1996 acquisition by Ruder Finn. Dr. Yosie has authored 40 publications including assessment of the chlorine hormone disruption debate, regulatory reform, risk assessment, and evaluations of corporate environmental management in the U.S. and developing nations. He is a frequent media commentator, speaks before many national and international groups, testifies before Congress, and is a member of the SAB's Integrated Risk Project, *Greenwire's* Board of Analysts, and the National Academy of Sciences Board on Environmental Studies and Toxicology. He received his doctorate in 1981 from Carnegie Mellon University.

Workshop Steering Committee, Plenary Speakers, and Discussion Groups

Steering Committee	
Larry Barnthouse (Workshop Chair)	LWB Environmental Services, Inc.
Greg Biddinger	Exxon Company USA
Bill Cooper	Michigan State University
Jim Gillett	Cornell University
Rick Haeuber	Sustainable Biosphere Initiative
Marge Holland	Ecological Society of America
Mike Slimak	U.S. Environmental Protection Agency
Terry Yosie	E. Bruce Harrison Co.

Plenary speakers	
Rick Haeuber	Sustainable Biosphere Initiative
Marge Holland	Ecological Society of America
Greg Biddinger	Exxon Corp.
Paul Portney	Resources for the Future
Bill Mulligan	Chevron Corp.
Joe Kovach	Cornell Agricultural Experiment Station
Dean Premo	White Water Associates
John Hartig	Wayne State University
Larry Harris	University of Florida
Sven-Olof Ryding	Swedish Association of Industries

Discussion group leaders	
Terry Yosie	Transboundary problems
Marge Holland	Landscape/watershed management
Jim Fava (Roy F. Weston Co.)	Product stewardship
Bill Cooper	Biodiversity protection and restoration
Jim Gillett	Urban environmental management
Greg Biddinger	Environmental restoration

Executive Summary

Environmental protection and management have been major international priorities for only about 25 years. From a strictly technical perspective, during the last decade it has become increasingly clear that the major environmental problems of the 1990s, such as regional environmental contamination, habitat destruction, and climate change, require interdisciplinary cooperation and new approaches to environmental risk assessment and management. The worldwide movement by both industrialized and developing nations toward balancing environmental quality, economic growth, and cultural development has greatly transformed and intensified the demands placed on ecologists. Success in achieving sustainable development will require advances in basic understanding of the interaction between human activities and ecological systems and innovative approaches to using ecological knowledge in environmental management. Scientists will have to step outside their traditional roles as academic researchers, resource managers, and regulatory specialists to participate much more actively in translating the intuitively appealing but operationally vague objective of "sustainable development" into concrete and achievable management plans.

The need for fundamental advances in the science underlying ecological resource management has been recognized by two major environmental societies, the Society of Environmental Toxicology and Chemistry (SETAC) and the Ecological Society of America (ESA). Leaders of both societies realized that the difficult technical and institutional challenges would require the efforts of many groups working together. Thus, a plan was set in place to jointly sponsor a workshop.

The workshop was held 25–31 August 1993. The format consisted of a combination of case study presentations and breakout groups, organized around six major themes:

- **Transboundary problems.** Issues such as transboundary flows of air and water pollution, shipment of hazardous and toxic waste from developed to developing countries, and control of water supplies serving multiple political jurisdictions have become important aspects of international relations. Transfer of environmental technology across international boundaries has been identified as a critical component of sustainable development.
- **Landscape and watershed management.** Many chronic environmental problems are more appropriately addressed at the watershed or landscape level than at the local level. Significant changes in both institutional arrangements and environmental management practices are required to effectively address environmental issues at this scale.
- **Product stewardship.** Many companies have recognized the need to integrate the principle of sustainable development into both their strategic planning and their day-to-day operations. Scientists working in industry must develop new technical tools and new ways of working within their organizations to ensure success.

- **Biodiversity protection and restoration.** Nearly all long-term human enterprises that involve extraction of goods and services from nature are benefited by, if not dependent on, biodiversity. Nearly all ecosystems have been directly or indirectly affected by human activities. Yet, less than 3% of the earth's surface is legally designated for biodiversity preservation, and only a fraction of these areas are secured from the continued loss of biodiversity. A wide range of scientific and institutional issues must be addressed if the environment is to be managed on a scale necessary to adequately protect biodiversity.
- **Urban environmental management.** Most residents of the United States and other developed nations now live in central cities and surrounding suburbs, so that resource consumption and control for all systems is dominated by demands and choices of these centers. Sustainability of the cities and their environs must be at the heart of the effort to develop an ecologically sound approach to global and regional environmental management.
- **Environmental restoration.** During the past two decades, there has been a rapid increase in our ability to restore natural ecosystems that have been disturbed by human activities. Successful restoration of ecosystems disrupted by resource extraction is dependent on adequate pre-disturbance planning and appropriate post-development restoration. A model is needed, applicable to both developing and developed nations, for incorporating sustainable environmental management (SEM) into the decision-making process of both governments and private companies.

Five case studies relevant to these themes were presented and discussed. Joseph Kovach discussed integrated pest management (IPM) and described ongoing work related to methods for demonstrating the effectiveness of IPM as compared to conventional chemical-intensive agriculture. John Hartig described a variety of local and regional initiatives undertaken to improve the ecological health and sustainability of the Great Lakes. Dean Premo described efforts of a paper company, working with a group of ecologists, to develop practical approaches to sustainable forest management. Sven-Olof Ryding described a technical tool, developed in a cooperative effort involving the Swedish government and industry, for evaluating the aggregate environmental impact of product manufacture and use, from resource extraction through final disposal. Larry Harris described the regional-scale ecological requirements for restoration of the Florida Everglades, the largest and most diverse wetland in North America.

Breakout groups organized around the six major themes discussed specific issues related to those themes and recommended positive steps that could be taken by environmental scientists to improve the use of scientific information to support SEM. Specific recommendations from each breakout group include these:
- Active technology transfer to developing nations, involving cross-functional teams to address specific needs such as industrial pollution control, energy efficiency, and forestry management. Such an initiative would have to be

participatory and would require full commitment and participation from the host country.
- Pursuit of regional-scale ecosystem management planning, formalized through intergovernmental Ecosystem Management Strategy Agreements (EMSAs). Such agreements already exist at the local and regional levels throughout the U.S., with the most prominent being the multistate Great Lakes Water Quality Initiative.
- Development by companies of internal environmental management systems to ensure that environmental objectives are considered along with other management objectives. Technical tools are important, but institutional commitment and organizational structures are the key to success. The Product Stewardship workgroup described a ten-step scheme for integrating an SEM philosophy into company operations.
- Establish and secure long-term commitments for protection of biodiversity. Neither economic markets nor conventional governmental decision-making processes are sufficient for this purpose because both operate on short time scales and respond more readily to short-term than to long-term concerns. Legally binding mechanisms must be found that insulate biodiversity protection from short-term political processes, while permitting long-term adjustments motivated by the right of each generation to make its own policy choices.
- Community-based, incentive-driven public–private partnerships provide the best approach to ensuring the sustainability of urban environments. The Remedial Action Plan (RAP) process developed for polluted rivers, harbors, and embayments on the Great Lakes is an example of such an initiative. Scientists have a key supporting role in providing the information and objective metrics for prioritization and performance.
- Command-and-control approaches to environmental restoration should be replaced by incentive-based approaches. Encouragement of stakeholder participation in the establishment of restoration goals provides an opportunity for greater flexibility in enhancing desirable ecosystem characteristics on a local or regional scale. A variety of technical tools and data sources are now available for helping scientists to aid stakeholders in establishing site-specific restoration objectives.

In the concluding session of the workshop, participants agreed on the following general conclusions:
- The appropriate management scale for most ecosystems is regional rather than local.
- The focus for management decision-making is likely to be regional and local rather than national.
- Feasible management approaches require involvement of all stakeholders; scientists, engineers, and economists cannot dictate solutions.

- Environmental management must be an adaptive process in which solutions are tried and modified based on experience.
- The private sector will play a major role in achieving environmental sustainability because 1) the private sector provides many opportunities for environmental improvement through pollution prevention, energy conservation, and improved product design, and 2) maintenance of social welfare requires an economically healthy private sector.
- Enhanced technology transfer is essential, both 1) from federal research and development agencies to state and local regulators and to the private sector and 2) from industrialized nations to developing nations.
- Education concerning the philosophy of and technical approaches to sustainable management is necessary; target audiences include both responsible decision-makers and the public at large.
- The rate of progress will be enhanced by increased communication among groups involved in the development and application of SEM plans.

Workshop participants recommended that SETAC and ESA should become actively involved in environmental decision-making. The joint program should include
- active outreach to states, communities, and businesses;
- educational programs to provide basic information to communities, state leaders, and business leaders; and
- involvement of scientists in developing countries, with the active support of the U.S. government or major private organizations.

Above all, scientists within both societies must understand that SEM is not a linear process in which experts prescribe solutions that then are enacted by governments or corporate executives, but rather a cyclical process in which scientists play a complex role that involves education as well as technical analysis and in which decisions are ultimately made by the people directly affected by them.

Abbreviations

CNET	Cornell Cooperative Extension Network
CFC	chlorofluorocarbon
CMA	Chemical Manufacturers Association
DFE	Design For the Environment
DFSD	Design For Sustainable Development
DOE	U.S. Department of Energy
EA	environmental audit/auditing
ECU	European currency unit
EIA	environmental impact assessment
EIQ	environmental impact quotient
ELU	environmental load unit
EMAP	USEPA's Environmental Monitoring and Assessment Program
EMSA	Ecosystem Management Strategy Agreement
EPS	Environmental Priority Strategies
ESA	Ecological Society of America
EXTOXNET	Extension Toxicology Network
FCA	full cost accounting
GIS	geographical information system
IPM	integrated pest management
LCA	life-cycle assessment
LDC	less developed country
MDNR	Michigan Department of Natural Resources
NAFTA	North American Free Trade Agreement
NEPA	National Environmental Policy Act
NOAA	National Oceanic and Atmospheric Administration

PCB	polychlorinated biphenyl
PLCA	product life-cycle assessment
RAP	remedial action plan
RCRA	Resource Conservation and Recovery Act
RMA	riparian management area
SBI	ESA's Sustainable Biosphere Initiative
SEM	sustainable environmental management
SETAC	Society of Environmental Toxicology and Chemistry
SIMPLE	Sustainability of Intensively Managed Populations in Lake Ecosystems
SMCRA	Surface Mine Reclamation Act
TEMS	Total Ecosystem Management Strategies
TQM	Total Quality Management
UNCED	United Nations Conference on the Environment and Development
USEPA	U.S. Environmental Protection Agency
USGS	U.S. Geological Survey
WSCAC	Water Supply Citizens Advisory Committee

Chapter 1

Introduction

As human populations expand, with concurrent demands for access to and/or use of natural resources for raw materials, food supply, housing, recreation, and other purposes, the global environment is increasingly taxed in its capacity to deliver all the goods and services demanded by humankind. Alterations of ecosystems will result in changes to the world as we know it. To lessen the rate and magnitude of change, it is imperative that global society embrace the concept of environmental stewardship. This requires that we manage our environment so as to ensure the continued supply of essential and desired environmental benefits. This idea is embedded in the term "sustainable development," as first used by the World Commission on Environment and Development (1987) and later adopted as a worldwide environmental goal at the United Nations Conference on the Environment and Development (UNCED). The fundamental premise of the Ecological Society of America's (ESA's) Sustainable Biosphere Initiative (SBI), which has been endorsed by the Society of Environmental Toxicology and Chemistry (SETAC), is that achievement of environmentally sustainable development will require advances in basic understanding of the interaction between human activities and ecological systems and innovative approaches to using ecological knowledge in practical, intentional environmental management.

The World Commission on Environment and Development (1987) and the UNCED defined sustainable development simply as the form of development or progress that "... meets the needs of the present without compromising the ability of future generations to meet their own needs." Sustainable development is an elegantly simple concept, but it means different things to different people and must be implemented in different ways in different societal and environmental contexts. Some (e.g., Ludwig et al. 1993) have suggested that published proposals for sustainable development are naive and doomed to failure. Without doubt, there are many uncertainties and questions about whether either the existing base of scientific knowledge or the ability of existing institutions to implement management actions is equal to the task. Nonetheless, the necessity for action is clear.

Even with the best possible institutional intentions, continuous scientific and technological innovation will be required to maintain human demands on the biosphere at levels that threaten neither the welfare of future generations nor the viability of ecosystems. The role of science includes the following:
- to develop new information and technologies that can be applied to preventing and/or solving problems;
- to identify ecological systems undergoing stress, or about to experience stress, and to devise management strategies to ameliorate this condition in a timely manner;

- to assist, through scientific analyses, in targeting priorities for action so that resources are applied to areas of need and to evaluate the effectiveness of remedial actions already taken; and
- to provide, together with other relevant disciplines (economics, engineering, etc.), scientific perspectives and methods through which decision-makers and citizens can be more fully informed about sustainability issues at local, regional, national, and global levels.

Because of the disciplinary orientation of the two sponsoring societies, the principal focus of this report is on 1) problems of reconciling environmental protection with economic growth and 2) problems of developed (North America, Western Europe) rather than developing nations. It is for this reason that the term "management" rather than "development" appears in the title of the workshop. For all intents and purposes, "sustainable environmental management" (SEM) as conceived by the workshop participants is sustainable development as applied to nations that are already highly developed and for which restoration of environmental resources degraded by past human activities is as important as preservation of resources during present and future activities.

Because of the enormous breadth of the subject matter, discussion at the workshop was limited to 1) characterizing differences between sustainable approaches to environmental management and conventional approaches as applied within the United States since the 1960s, 2) defining certain general principles, and 3) discussing six environmental management topics that are of major current interest. The principal definitions and concepts involved in SEM, as developed by workshop participants, are presented below.

Sustainable environmental management has three principal components: environmental protection, economic health, and social responsibility (Figure 1).

Figure 1 Principal components of sustainable environmental management

Environmental protection occurs as a result of utilizing scientific approaches, systems, and tools to identify, implement, and sustain environmental improvements. Environmental areas of concern include ecological health, human health, and resource depletion. Economic health is achieved by ensuring that material, financial, and cultural resources are efficiently and ethically used to improve the vitality of institutions and the quality of life; this requires that environmental protection measures and economic development policies complement rather than conflict with each other. Achieving social responsibility requires the acknowledgment by all institutions, both public and private, that their actions affect the well-being of society in many ways. Socially responsible institutions work to ensure that in the course of pursuing their objectives (e.g., economic gain), the direct and indirect effects of their activities are fully accounted for and, if necessary, ameliorated. The interconnectedness among societal values and needs, environmental improvement, and economic health is a cornerstone of the concept of sustainable development. A comprehensive description of types of societal impacts that are relevant to SEM was developed at the February 1992 SETAC workshop on LCA Impact Assessment (Fava et al. 1993) and is presented in Table 1.

Table 1 Social welfare impact categories

Impact category	Potential impact	Impact category	Potential impact
Demographic	Fertility and mortality	**Community**	Community conflict
	Morbidity		Community cohesion
	Migration		Land use
Economic	Opportunity and transaction costs		Services and facilities
	Real property value		Health
	Inflation		Welfare
	Growth		Education
	Sectoral (fishing, recreation, tourism, etc.)		Public safety
Fiscal	Public services and facilities		Community infrastructure
	Supply/demand		Housing
	Costs/revenues		Transportation
Sociopolitical	Legal		Physical appearance
	Governmental		Community identification
	Credibility		Community satisfaction
	Centralization		Community institutions
	Intergovernmental relations		Labor force availability and participation
	Regulatory		
Social	Social integration and cohesion	**Family**	Family structure
	Higher principles		Employment
	Social networks and social support		Family stability
	Social mobilization and participation	**Sociocultural**	Way of life
	Social conflict and tension		Cultural survival
	Social disorganization and deviance		Cultural heritage
	Autonomy and dependency		World view
			Values
			Higher principles
			Social justice
			Aesthetics
			Environmental values
		Psychosocial	Self-esteem
			Autonomy and dependency
			Apathy and alienation
			Uncertainty
			Anxiety
			Stress
			Stigma
			Psychopathology

Source: Fava et al. 1993.

Workshop participants identified three characteristics that, from an environmental perspective, distinguish sustainable development and management strategies from conventional strategies: scale, complexity, and inclusiveness. By "scale" is meant the space and time dimensions encompassed by management actions. Sustainable management strategies often must be implemented over much larger areas and over longer time periods than strategies that have been adopted in the past. The cases of Everglades environmental management and Great Lakes management presented at the workshop are examples. Ecological functions in the Everglades ecosystem cannot be preserved unless the protected region of South Florida is expanded to include the rivers draining into and out of Lake Okeechobee. Similarly, the states bordering the Great Lakes have been forced to develop basinwide management plans to enhance salmon production and reduce toxic chemical inputs.

By "complexity" is meant the number of resources and human uses that must be simultaneously managed. Because sustainability-based strategies focus on the quality of the resource rather than on the type or intensity of a particular stress, all of the human influences on each resource and all of the services provided by that resource must (at least in principle) be considered. Sustainability-based management of forests, for example, requires the development of harvesting strategies that maintain the ability of the forest to provide necessary habitat for native biota. Product stewardship strategies now being implemented in the United States and Europe require the evaluation of materials requirements, environmental impacts, and waste management implications for all stages of the product life cycle, from resource extraction to ultimate disposal.

By "inclusiveness" is meant the inclusion of human populations and institutions as integral components of the management process. Walters (1986) noted that fisheries management is really a process of managing people, not of managing fish. Perhaps the greatest difference between sustainability-based and conventional approaches to environmental management is the recognition that society is a part of the solution as well as the source of the problem. There is ample evidence of the validity of this conclusion, both in the published literature and in the materials discussed at the workshop. The product stewardship movement, a private-sector initiative discussed extensively in Chapters 2 and 4, has been endorsed by chief executive officers of dozens of major international corporations as a means of reducing environmental impacts of industrial activities while maintaining or increasing economic efficiency (Schmidheiny 1992). SETAC's Life-Cycle Assessment Advisory Group has fifteen members representing fourteen corporations, government agencies, and educational institutions in North America and Europe. Regulatory approaches to environmental management are rapidly moving away from exclusively nationally mandated standards ("command-and-control" strategies) to economic incentives, locally or regionally based regulations, and voluntary compliance. The importance of including the people affected by environmental decisions in the decision-making process was emphasized by every discussion group; a substantial fraction of the text of Chapter 4 is devoted to sociopolitical aspects of environmental management.

Figure 2 depicts a sequence of steps leading to the development of SEM strategies, as developed at the workshop. The starting point for any sustainable management plan is identification of the environmental resources and services to be protected and managed. These may include biotic diversity, water quality/quantity, commercial resource products (fisheries and forests), recreational or aesthetic values, or any other aspect of the environment valued by society. The identification of these resources and services is necessarily a sociopolitical process involving resource scientists; private companies; local, state, and federal governments; and citizens. Human perceptions and values, as expressed through the democratic political process, determine management goals and objectives. These goals are expressed in very general terms, such as "preserve rare and diverse ecosystems like the Everglades," or "prevent deterioration of air and water quality."

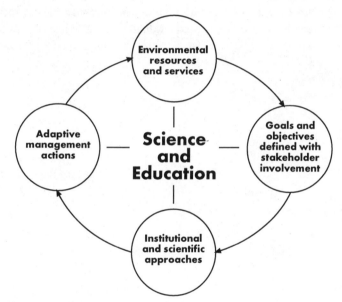

Figure 2 Development of sustainable environmental management strategies

The central operational task of SEM is to transform society's goals into effective management actions that actually sustain environmental resources and services. In the past, many scientists have treated this task as a purely technical activity. We now recognize that scientists alone cannot create viable management approaches. Both institutional approaches and scientific approaches are required. Institutional approaches include product stewardship, economic incentives, local and regional management plans, and other supplements to conventional command-and-control environmental standards. The disciplines needed to develop adequate scientific approaches include engineering, economics, and other social sciences, in addition to the natural science disciplines (ecology, environmental toxicology, chemistry, etc.) that have been the traditional modalities of environmental management.

Development of effective actions requires that institutional approaches and scientific approaches be matched in terms of resources and services addressed, spatial extent of managed ecosystems, and time scales required for effective implementation. Because of the many uncertainties, both scientific and sociopolitical, that affect our ability to manage environmental resources, the effectiveness of actions must be continually monitored and adjusted. Holling (1978) and Walters (1986), who have long advocated experimental and iterative approaches to environmental management, have termed this process "adaptive management." Understanding of the adaptive nature of SEM systems leads immediately to a recognition that environmental problems will not suddenly be solved once scientists have found the right answers and managers have instituted appropriate regulations. The process will, on the contrary, be an evolutionary one in which management actions are implemented, monitored, evaluated, modified, and, if necessary, replaced.

Chapter 2

Issues in Sustainable Environmental Management

Environmental sustainability is a general principle that affects every aspect of the interaction between human societies and their environment. The steering committee chose to focus on a small number of society–environment interactions for which 1) the benefits of a sustainability-based approach can be clearly demonstrated, 2) the necessary technical tools are available, and 3) evidence of shifts to more SEM approaches already exists. The hope of the committee was that exploration of these specific cases would provide conclusions and recommendations relevant to a much broader array of sustainability-related topics. At a planning meeting in November 1992, six suitable topics were identified:
1) transboundary problems,
2) landscape and watershed management,
3) product stewardship,
4) biodiversity protection and restoration,
5) urban environmental management, and
6) environmental restoration.

The six do not exhaust the possible elements of SEM, but they are representative of the range of problems confronting environmental decision-makers today. Most reflect trends in environmental management that began prior to the workshop and have continued to accelerate since the workshop: development of regional environmental management goals, transfer of regulatory authority from the federal government to states and local communities, private-sector leadership, and global environmental technology transfer. For each topic, a discussion group at the workshop was charged with defining the scope of the topic, summarizing current approaches, and identifying emerging approaches consistent with a sustainable management philosophy. For clarity of presentation, the discussion summaries are separated into this chapter (problems) and Chapter 4 (new approaches) of the report. The case studies commissioned for the workshop are presented in Chapter 3.

Transboundary Problems

In recent years, transboundary environmental issues (e.g., flows of air and water pollution, hazardous wastes, toxic substances, and even increased flooding) have become a focal point in discussions of SEM. Increasingly, they form an important set of issues in international relations and, in many cases, a growing source of international tension. Internationally, transboundary issues may involve various combinations of parties —

e.g., flows of air or water pollution between developed countries; from a less developed country (LDC) to a developed one, or vice versa; or between LDCs. Examples include these:
- Emissions from power plants in the Midwestern region of the United States and their potential role in acid deposition in Canada have long been a sore point in U.S.–Canadian diplomatic relations.
- Similar concerns about transboundary flows of air and water pollution, in this case from Mexico to the U.S., pose a major roadblock to successful implementation of the North American Free Trade Agreement (NAFTA).
- Concern over flows of hazardous and toxic wastes from developed countries to less developed countries is a growing source of tension in north–south hemisphere relations, as attested to by recent cases of barges with industrial waste cargo being denied docking privileges.
- Over the last ten to twenty years, the increasing frequency and severity of flooding in the Ganges Plain of India has been attributed to deforestation in the Himalayas, some occurring in India, but much of which is traced to Nepalese forest practices.

The context of transboundary environmental issues is not confined to relations among countries but may involve jurisdictional boundaries within a single country as well. As with the U.S.–Canada example, the case is frequently made that acid deposition in lakes and streams of the northeast U.S. can be traced to emissions from midwestern power plants. Similarly, high levels of industrial wastes and emissions of all types have created deteriorating environmental conditions within the borders of countries throughout all of Eastern Europe.

Addressing such issues and achieving sustainability will depend upon a process involving at least three basic steps:
1) Create an information base sufficient to understanding past and present conditions and future threats to physical and natural resources. This information base will entail the development of new knowledge through research and monitoring, but it must also involve synthesis and integration of existing knowledge in order to make it more accessible.
2) Develop new strategies and techniques to manage natural and physical resources and productive processes, based upon both new and existing information.
3) Ensure rapid transfer of management systems and technical tools to people who need them for immediate problem-solving.

Transboundary environmental issues involve all economic sectors, including production and manufacturing of all types, mining and minerals extraction, energy (not only issues of extraction and refining but also questions of energy efficiency), natural resources management, and even agriculture. The evolution of U.S. environmental experience over the past two decades has produced a body of data, management tools, and technologies designed to deal with such issues. The products of these years of experience are readily adaptable for application to many regions within the U.S. as well as to other nations, to improve their environmental quality. One of the most pressing needs at this time is the transfer of these data and management tools to industries, localities, and nations that could benefit from that accumulated knowledge and experience. To be successful, the technology transfer process must respect local cultures and values.

Landscape and Watershed Management

The management of landscapes and watersheds is best served through the use of natural rather than political boundaries in setting management goals. Management strategies need to be developed that utilize watershed boundaries to consider a drainage basin as a whole, or regions to consider regional processes such as acid deposition. Managers, planners, policy makers, and scientists need to look at natural resource problems from a larger-scale perspective than traditionally has been used. There is a need for a landscape perspective in conducting research, as well as in managing ecological systems (Holland 1993).

Many chronic environmental problems are more appropriately addressed at the watershed or landscape level than at the local level. Examples of such problems include these:
- nonpoint source pollution, including agricultural chemicals, hazardous waste sites (localized but often numerous), atmospheric deposition, urban runoff, and contaminated sediments;
- habitat loss or destruction;
- undesirable nonindigenous species;
- point source pollution, including inadequate or outdated sewage treatment facilities, construction sites, and marinas;
- modified hydrology, including stream diversions, channelization (levees), impoundment (dams), reduced infiltration (increased variation in runoff), and shoreline erosion; and
- interruption of ranges or migration routes.

Ideally, these problems should be solved without significant disruption of important human activities such as agricultural production, water distribution (drinking, irrigation, bathing), housing or habitation, industry, transportation, forestry, fisheries, recreation, and perception of environmental quality or aesthetics.

Six properties of functional ecosystems are important to recognize. These properties may vary in degree from one ecosystem to another, but they are all essential to the maintenance of functioning ecosystems:

1) adequate water, soil (sediment), and air quality and quantity (including hydroperiods);
2) diversity and adequate quantity of habitat types;
3) viable populations of diverse native species (maintain gene pool);
4) resilience (structural and functional) and regenerative ability;
5) self-sustainability, (minimum subsidy, minimal ecological management); and
6) connectivity.

Resilience and connectivity are especially important concepts for maintenance of sustainable landscapes and watersheds. "Resilience" refers to the ability of a system to respond to stress; if a system is able to return in a timely fashion to its original characteristics after some form of disturbance, then it is said to be resilient. "Connectivity" refers to the functional relationships in natural systems, such as the close association of a stream with its flood plain. A loss of connectivity may occur directly, through diking and channelization, or indirectly, through the regulation of flood regimes via navigation dams (Pacific Rivers Council 1993). In the case of terrestrial systems, connective corridors are believed essential to allow animals and plants to disperse and to maintain viable populations in increasingly fragmented ranges.

Environmental management requires that effects of management actions be predicted. However, ecologists have not come very far in developing a predictive science. There are always scientific uncertainties related to predicting what will happen in the environment, given the complexity of the environment and our limited understanding of all the interactions. As ecologists, environmental toxicologists, and natural resource economists have been called upon to help solve environmental problems, the need for improved predictive capability has become more apparent. Limitations on predictive capabilities have impeded the acceptance and implementation of ecological concepts.

At the present time, given our collective understanding of ecological processes and ecosystem functioning, predictions concerning the resistance and resilience of specific ecosystems are usually highly uncertain. It is also very difficult to develop operational definitions of connectivity and to predict how changes in connectivity will affect the structure and function of ecosystems. Landscape patterns have an impact on biodiversity, but the exact relationships are uncertain. Ecologists have established, e.g., that forest fragmentation has contributed to a decline in the abundance of spotted owls, but predictive models that could aid in managing these forests to provide both timber production and spotted owl habitat do not yet exist. In short, both institutional barriers and scientific barriers currently impede the development of rational landscape and watershed management strategies.

Product Stewardship

Many companies recognize the importance of integrating product stewardship (ensuring compatibility between a product and the environment) into their strategic planning and their day-to-day operations. There is ample evidence that enhancing the quality of the current environment as well as establishing policies to sustain the environment must take on greater priority within this generation and those that follow. While this belief is commonly held by governments, environmental groups, industry, academics, and the general public, there is considerable debate on how to achieve improved environmental conditions without sacrificing current economic benefits of environmental stewardship.

In some cases, integration of SEM is limited to a gate-to-gate perspective, with product and process design based on traditional considerations of function and cost, without direct — and in many cases, indirect — consideration or integration of environmental protection ("gate-to-gate" indicates the time beginning when raw materials are purchased by a manufacturer and ending when the resulting product is sold). Table 2 organizes current health and environmental welfare issues in terms of their function in product design and manufacture. The majority of issues are driven primarily by economic health, with fewer driven by environmental protection. Social responsibility influences only two of these issues.

Table 2
Current integration of health and environmental issues within typical manufacturing companies

	Issue		
Function	Social responsibility	Economic health	Environmental protection
Marketing/sales		X	
Research and development		X	X
Manufacturing		X	X
Compliance		X	X
Government affairs			X
Finance		X	
Procurement		X	
Human resources	X	X	
Legal		X	X
Public relations	X	X	X
Business management		X	X

One distinction that sheds light on the difference between traditional business practices and the new SEM way of thinking concerns the way in which these alternative approaches address needs and problems presented by clients. Currently, organizations attempt to serve their clients by addressing single, separate issues in an isolated fashion. Clients' needs are met sequentially, item by item, and solutions are sought from a narrow perspective with little regard to ecological and social implications. Often, environmental and social issues are addressed well after large investments in technology, organizational change, or political capital have been made. In contrast, an organization that practices SEM identifies and brings a wider set of stakeholders into the decision-making process, serving clients' needs and advancing the organization, while directly addressing the ecological and social implications of its actions at the outset.

Some of the more progressive companies, agencies, and academic departments have begun to integrate single-issue SEM concepts and tools such as product stewardship (e.g., chlorofluorocarbon [CFC] reduction and handling of solid waste) into their product and process development schemes. This integration is primarily due to market pressures such as federal procurement policy, ecolabeling, and supplier selection based on environmental criteria. The chemical industry, through its Chemical Manufacturers Association (CMA), established a Responsible Care program that applies a series of principles and practices to manage the risk of chemicals in the environment. One of the management codes identifies practices that enable a company to develop and implement proactive product stewardship programs. Many chemical companies have signed onto Responsible Care programs but still are struggling with how to implement such far-reaching policies and integrate the corresponding thought processes and activities into a new way of life.

Although these activities are clear signs of progress in incorporating the philosophy of SEM into business decision-making, coordination among groups interested in fostering and integrating SEM has been poor. Academic involvement has been limited, and adversarial relationships have impeded cooperation among industry groups such as CMA, governmental agencies, and nongovernmental organizations.

To assist in overcoming these debates, environmental professionals from all sectors are developing tools (e.g., life-cycle assessment [LCA], Design For the Environment [DFE], etc.) that represent a major step forward in integrating environmental considerations across a spectrum of possible environmental releases from "cradle to grave." However, as powerful as these new developments appear to be, they do not ensure that environmental improvement, economic considerations, and social responsibility will be properly balanced. What is needed is a concept and ultimately a process that will help society make decisions regarding the best balance of environmental needs, economic realities, and societal wishes. Sustainable environmental management, a concept that attempts to integrate all three considerations, provides a powerful new way of thinking in regards to environmental health for everyone.

Incorporating SEM principles into business decision-making will require commitment and education at all levels of an organization, including stockholders. Not unlike their experience in implementing Total Quality Management (TQM) systems or other major programs, organizations will face the need to explain another change in the status quo. A clear statement of principles that can be understood at all levels of business organizations and that can be embraced by government agencies and nongovernmental organizations as well would greatly facilitate this necessary educational process. Such a statement would also serve to reduce the level of confrontation among these three societal sectors when they deal with environmental issues.

Biodiversity Protection and Restoration

Building a case for biodiversity conservation as a premise for achieving a sustainable environment is not difficult, inasmuch as nearly all long-term human enterprises that involve extraction of goods and services from nature are benefited by, if not dependent upon, biodiversity. Not only does biodiversity constitute the very fabric of renewable natural resources, but it also promotes the processes through which these resources are regenerated.

Nearly all ecosystems have been directly or indirectly affected by human activities. Some of these ecosystems play a role in maintaining valued environmental services that are disproportionate to the area they occupy. In the western United States, riparian woodlands contain the greatest abundance and diversity of species, yet they occupy less than 1% of the terrestrial landscape. Similarly, freshwater habitats throughout the world support a high diversity of aquatic and terrestrial species, while (with a few exceptions such as the Great Lakes) occupying relatively small areas.

Tropical rain forests, estuaries, coral reefs, and similar systems that occupy much larger areas and contain some of the greatest biodiversity would seemingly be at less risk than those occupying small areas. However, because of large-scale exploitation (e.g., timber harvesting and land clearing in rain forests) and exposure to the indirect effects of human activities (e.g., the accumulation of toxic substances in estuaries through point source and nonpoint source pollution and the sedimentation of coral reefs that results from unmitigated land and resource use), these systems are also at risk. At current rates and under present conditions of development, some areas may survive for a few more decades, but others are already gone.

Less than 3% of the earth's surface is legally designated for biodiversity preservation (i.e., parks and preserves), and only a fraction of these areas is secured from the continued loss of biodiversity. Under conditions of depleted biodiversity, ever-increasing proportions of the fundamental functions characteristic of forest, range, and agricultural landscapes will require technological intervention. This steady conversion of coevolved systems to human-managed systems undoubtedly increases both the economic costs of natural resource management and the risks that management failures will adversely (perhaps disastrously) reduce environmental and human welfare.

When one considers the value of clean water and air, the importance of natural ecosystems becomes evident. The turnover of water, gases, and organic matter in natural areas by natural processes is overwhelmingly responsible for cleansing air and water of the pollution created by humans. Predation, competition, and natural mutation are the fundamental functions underlying the maintenance and generation of further, as of yet unconceived, biodiversity. Conversely, just as organisms are in part controlled by environment, they are also capable of altering ecological communities, ecosystems, and even landscapes. Consequently, biodiversity "back-feeds" (via feedback loops) itself, inducing habitat variability and more diversification at both lower and higher levels of organization or scale.

Both the term and the concept of biodiversity connote tangible entities such as genes, phenotypes, species, and ecosystems. Although these entities are critically important in their own right, the predominance of goods and services that humans require from biodiversity derives from the active metabolism and interaction of large numbers or spatial expanses of these entities, i.e., the work that they do. Examples of these services are water and air purification, hydroperiod regulation, disease and pest control, climate amelioration, etc. We coin the term and concept of "ecodiversity," albeit more prosaic and utilitarian, because it is more relevant to sustainability-based environmental management than is the term "biodiversity."

Agriculture, an activity that spans millennia, is the most apparent of all human management activities. In spite of the availability of many hard-technology options that minimize the dependence of crop production on natural processes and resources, biodiversity is playing an increasingly important role in modern agriculture. Biodiversity is the very basis of Integrated Pest Management (IPM) strategies intended to reduce the use of chemical pesticides. Biodiversity furnishes the predators, parasites, pathogens, and corresponding biotic interactions responsible for the control of damaging organisms. A host of other natural processes such as biological fixation of atmospheric nitrogen via bacteria–plant symbioses are critically important complements to technology-based farming systems. Improved absorption and conservation of nutrients, especially phosphorus, via mycorrhizal associations with agricultural and forest plants is critical to high-production systems. Soil fertilization, physical amelioration, and improvement of water retention through organic matter decomposition by a plethora of soil organisms are essential for the maintenance of the productive capacity of terrestrial ecosystems. Natural populations act as stocks of genetic material necessary for the development of productive, efficient, pest-resistant domestic organisms. Attractive and repellent plants are important to insect pest management. These are but a few of the services rendered by natural, integrated biodiversity complexes in agriculture and forestry.

A range of issues must be addressed in development of any plans for protection of diversity:

- Existing resource conservation paradigms are arguably inadequate to support even present levels of human population and global economic activity into the future.
- Issues of scale, both temporal and spatial, are now known by ecologists to be more important than previously thought. It is easy to document that precisely opposite ecological relations exist between principal variables depending upon the scale of investigation.
- The purposeful transfer of thousands of species among continents to date has served to homogenize the earth when viewed from the level of the biosphere.
- Migratory species are vulnerable to the loss of migration corridors or habitat. Protection of these species is especially difficult because habitats occupied during different seasons or life stages may be separated by thousands of miles and may be affected by unrelated natural and anthropogenic influences.
- Contextual setting is likely to be as important in developing and maintaining effective biodiversity refugia as is the mapping of biodiversity "hot spots" and establishment of preserves simply to contain said resources. In more colloquial terms, how you manage the landscape surrounding a park or biodiversity preserve may be more important than how or whether you manage the reserve itself.
- Ecosystems are dynamic systems. Episodic natural disturbances, both physical (e.g., hurricanes, droughts and tornadoes) and biological (e.g., population outbreaks), cause ecosystems to be dynamic. An understanding of the natural disturbance phenomena in various regions of the earth — including their frequency, intensity, spatial distribution, and extent — is critical to effective sustainability-based management of biodiversity.
- Ways to achieve and reward long-term institutional and individual commitments must be developed in concert with trust between private, corporate, and governmental entities in order to accomplish sustainability-based stewardship.
- The dilemma of ownership rights is a critical item of debate in aligning existing property rights laws with sustainability-based stewardship for biodiversity. Conflicts will arise between property owners and governmental entities. Resolution of such conflicts will depend upon development of suitable decision-making processes, including establishment of appropriate guidelines for selecting potential biodiversity areas and compensating property owners and for creating opportunities to negotiate with affected parties over site-specific issues.
- The relative effectiveness of alternative biodiversity conservation strategies must be evaluated. To date, two diverse schools of thought have dominated biodiversity conservation strategies: the set-aside, leave-it-alone approach as manifested by parks and preserves, and the managed, sustained-yield approach that has dominated forestry and fish and game management.

- The distribution of profits derived from direct marketing of biodiversity entities and the development of new consumer products and/or valuable pharmaceutical products must be equitably resolved.
- Existing institutional programs and policies that encourage and/or sanction the introduction of exotic species and genetic materials need to be scrutinized in the context of sustainability-based stewardship for biodiversity. Exotic introductions that might seem to increase biodiversity at a local scale or over a short time scale may be found to decrease biodiversity when evaluated at the regional scale or over a longer time scale.

Urban Environmental Management

Most residents of the United States and other developed nations now live in central cities and surrounding suburbs, so that resource consumption and control for all systems is dominated by demands and choices of these centers. Sustainability of the cities and their environs must be at the heart of any effort to develop an ecologically sound approach to global and regional environmental management. To date there has been little recognition of the close interconnections between economic and ecological processes and human ecology within urbanized environments (Gillett et al. 1992). Important aspects of urban life (e.g., urbanity and civilization and their juxtaposition with concepts of wilderness and harmony) are emergent properties that transcend the physical and biological features of ecosystems and landscapes (Slocombe 1993).

Cities and their suburbs combine elements of every sector of concern for sustainable management of the environment. Continual improvement (redevelopment) is needed to permit cities to retain (regain) livability by providing transportation, safety, absence of toxic chemicals, economic security, health care, and housing. Given that urban environments per se are highly modified, it is unrealistic to expect city or suburban land use to return to some pristine state.

The rising expectations that attract people to cities and hold residents there involves assumptions about the character of a "healthy city," which the World Health Organization defines as

> one that is continually developing those public policies and creating those physical and social environments which enable its people to mutually support each other in carrying out all the functions of life and achieving their full potential (Healthy Toronto 2000 Subcommittee 1988).

The city brings in energy, nutrients, and diverse biota; restructures the land- and waterscapes to meet its physical, social, and aesthetic needs; and discharges wastes into air, water, and soil. Such parallels to organismic activities as metabolism and growth typically generate anthropomorphic symbols (the "heart" of the city, the "arteries" of transportation) and ecological comparisons with other systems. Landscape ecology arises directly from that perspective.

Toward which characteristics of a healthy city should we aspire? For metropolitan Toronto, Ontario, these have been stated as "...clean, green, usable, diverse, open, accessible, connected, affordable, and attractive" (Crombie 1990). To this list can be added

- a sense of overall community (as well as subcommunities or neighborhoods that exist within the greater whole),
- respect for cultural and racial diversity, and
- freedom from violence.

"Usable" and "attractive" imply public safety and environmental health, but "usable, accessible, connected, and affordable" may also speak to transportation, housing, and economic viability. By way of contrast, an unhealthy city is one where anxiety (due to concerns about safety, pollution, lack of transit options, etc.) leads people to reconsider whether they really have to go there. It is an unhealthy city that gives up the most revenue (through subsidies, tax reductions, e.g.) to attract new industry and that may be forced to use raw political, economic, or social power to impose its will on neighboring areas and the distant sites of its resources.

Both ancient and recent history underline the dynamics of cities. Within cities, neighborhoods may fall into decay, only to be redeveloped and gentrified. Urban systems span wide ranges of time and draw upon resources from vast distances. Many cities (or sites) have had continuous occupancy for centuries, although the rise and fall of neighborhoods may have time constants of decades. Other cities are as fast-growing as a malignancy, radically and perhaps even terminally changing almost before our eyes. The explosive growth and inability to handle wastes in air or water, to provide useful employment and proper housing, and to deal with the health and education of adults and children put these cities at serious risk. Seldom will any one ecological or environmental factor be involved.

The past 20 years have engendered efforts to bring wastes, energy and other resource use, and human population under control, but these efforts have not been adequate. Legislation and regulation have moved to protect species diversity and vital habitats, to address abandoned hazardous wastes, and to reduce nonrenewable resource consumption. All of these measures can be included in "conventional" environmental management, which may be described as a top-down, command-and-control, medium-by-medium approach (e.g., air, water, land). Although potentially effective in some instances, the conventional methods are apt to be piecemeal, arbitrary, and inadequately integrated among themselves and with other management actions. Appropriate external controls (e.g., input from stakeholders) and ties to critical functions, such as managed population growth, are often absent. New approaches are needed to promote sustainable maintenance of human welfare in urban environments.

Environmental Restoration

During the past two decades, there has been a rapid increase in our ability to restore natural ecosystems that have been disturbed by human activities. The primary motivation for this progress has been the realization that ecosystems disrupted by resource extraction activities can be returned to productive and diverse natural states only with the aid of 1) pre-disturbance planning to reduce adverse impacts and 2) substantial post-development restoration efforts. The realization that prior restoration planning is critical has resulted in the passage of legislation such as the National Environmental Policy Act (NEPA 1970) and Surface Mining Control and Reclamation Act (SMCRA 1977) in the United States and of similar laws in other countries. Considerable efforts have been made to develop methods and implementation approaches for restoring terrestrial and aquatic ecosystems disturbed by human development activities. Progress has come from university research programs and from industry's application of new restoration and monitoring practices.

The continuing increase in human population and consequent increased demands for living space and food production have resulted in additional disturbance of ecosystems. Irretrievable losses of biodiversity have already occurred on a global scale. Any concept of ecosystem sustainability should, therefore, include a combination of conservation and restoration practices.

In developed countries such as the United States, the drivers for restoration are command–control mechanisms in the form of federal and state regulations that place rigid requirements on the goals and methods of restoration. Under these mechanisms, restoration goals generally focus on ecological baseline conditions within the affected area and do not consider the diversity and proportion of similar habitats and ecosystems within a region. Although this is arguably a desirable restoration goal, it restricts the opportunities for flexible restoration planning that could perpetuate the natural diversity of habitats and produce a more sustainable mix of habitats and species over broader geographic areas. In the face of increasing land-use pressures and disproportionate losses of ecologically important habitats, greater flexibility in restoration planning is necessary.

The situation in developing countries differs from that in developed countries in that resource development is still increasing. However, governments often have neither the financial resources nor the infrastructure to implement effective restoration policy. The result is an uneven mix of disturbance activities that range from state-of-the-art, environmentally responsive projects to regional development schemes that exploit and frequently waste available resources for purposes of short-term gains, while ignoring long-term consequences. Economic incentives for resource development exist universally, but these incentives generally do not include considerations of environmental costs of development and exploitation. A model is needed, applicable to both developing and developed nations, for incorporating SEM into decision-making processes. Application of the model must, however, be flexible enough to accommodate varied cultural and economic circumstances.

Chapter 3

Case Studies

Integrated Pest Management
(Joseph Kovach)

In general, agricultural areas represent biodiversity deserts in relation to the original local biota. Additionally, the majority of organisms present, comprising the crop plants, their competitors (weeds), consumers (herbivores), parasites and other closely related organisms (weed-related herbivores, secondary consumers, etc.), normally are not native.

This means that agriculture in itself is based on a very narrow biological blend, comprised essentially of exotic species selected specifically for productivity in a controlled environment and of invading species that are generally preadapted to the very energy- and resource-intensive type of ecosystem. This simple community structure is responsible for the likelihood that, at any given time, at least one of these estranged populations will not be in check. When this unchecked population happens to endanger the crop, external control measures are warranted.

At present, there are effectively two technically and economically important interventions available for such pest control problems. The most commonly employed one is the prescriptive spraying of pesticides, which normally results in further depletion of biodiversity within the ecosystem and other affected areas. Obviously, this is not an SEM strategy.

A second and increasingly important type of intervention is biological control. It consists of repeated releases or definitive establishment of specific natural enemies of the species whose population is threatening the crop in that given instance. Such an intervention does result in increased biodiversity, and in cases in which the natural enemy manages to survive in the system in the long term, it certainly enhances sustainability. However, natural environmental oscillations, changes in management practices, or substitution of the crop system altogether may lead to the emergence of other damaging pest species to damage levels, requiring new external interventions. It becomes quite obvious that large-scale modern agriculture, as it is practiced today, is not a sustainable activity when pest management is considered.

On the other hand, examples of traditional sustainable agriculture abound. South Asian rice paddies, Mexican chiampas, Central American *frijol tapado* (covered beans), and some instances of slash-and-burn shifting agriculture, to enumerate only a few examples, have been practiced in the same pieces of land for generations, with minimum inputs besides labor. The lesson is that large-scale modern agriculture must evolve to more diversified agroecosystems less dependent on external inputs. Developing the engineering, equipment, and agricultural science to accomplish this goal may be considered the grandest challenge for the agricultural sector in the future.

In recent years, public concern about food safety, groundwater contamination, and other environmental problems has caused the agriculture industry to pay increased attention to IPM and other approaches to reducing pesticide use. IPM is a pest management strategy that uses biological and cultural pest management methods to permit production of the same (or better) quantity and quality of agricultural products with smaller quantities of chemical pesticides compared to other, more conventional strategies. In past IPM programs, pesticides generally were chosen based on their efficacy or cost rather than on their potential environmental impact. No formal methods were available to assist growers in selecting pesticides that minimized impacts either on beneficial natural enemies or on the persons applying the pesticides.

Because of the U.S. Environmental Protection Agency's (USEPA's) pesticide registration process, a wealth of toxicological and environmental impact data is available for most pesticides commonly used in agricultural systems. To assist growers and other IPM practitioners in making more environmentally sound pesticide choices, the Cornell University IPM Program has organized and published this information in a usable form (Kovach et al. 1992). A method was developed for calculating the environmental impacts of most common fruit and vegetable pesticides (insecticides, acaricides, fungicides, and herbicides) used in commercial agriculture. The values obtained from these calculations can be used to compare the environmental impacts of different pesticides and pest management programs.

The Extension Toxicology Network (EXTOXNET), a collaborative education project between the environmental toxicology and pesticide education departments of Cornell University, Michigan State University, Oregon State University, and the University of California, was the primary source used in developing this database (Hotchkiss et al. 1989). EXTOXNET contains pesticide-related information on the health and environmental effects of approximately 100 pesticides. A second source of information used was CHEM-NEWS of the Cornell Cooperative Extension Network (CNET). CHEM-NEWS contains approximately 310 USEPA Pesticide Fact Sheets that describe the health, ecological, and environmental effects of pesticides reviewed for reregistration (Smith and Barnard 1992).

The impact of pesticides on arthropod natural enemies was determined using the SELCTV database developed at Oregon State (Theiling and Croft 1988). These authors rated the effects of about 400 chemical pesticides on over 600 species of arthropod natural enemies. Published data on leaching, surface loss potentials, soil half-life, and bee toxicity are also included in the database.

The above information was used to develop an environmental impact quotient (EIQ) that addressed common concerns expressed by farm workers, consumers, pest management practitioners, and other environmentalists interested in minimizing the health and ecological impacts of pesticide use. A detailed explanation of the EIQ equation and values of EIQs for more than 100 pesticides are presented by Kovach et al. (1992).

A simple equation called the "EIQ field use rating" accounts for variations in pesticide formulations and use patterns:

EIQ field use rating = EIQ × % active ingredient × rate.

This adjustment permits comparisons of environmental impacts between pesticides and pest management programs. Table 3 shows the theoretical environmental impacts of several different pest management approaches that have been used in research projects to grow Red Delicious apples in New York.

Table 3 Theoretical environmental impact of pest management strategies for Red Delicious apples in New York

Material	Conventional pest management strategy				
	EIQ	ai[1]	Dose	Applications	Total
Rubigan EC	27.3	0.12	0.6	4	8
Captan 50WP	28.6	0.50	3.0	6	257
Lorsban 50WP	52.8	0.50	3.0	2	158
Thiodan 50WP	40.5	0.50	3.0	1	61
Guthion 35WP	43.1	0.35	2.2	2	66
Cygon 4E	74.0	0.43	2.0	3	191
Omite 6EC	42.7	0.68	2.0	2	116
Kelthane 35WP	29.9	0.35	4.5	1	47
Sevin 50WP	22.6	0.50	1.0	3	34
Total environmental impact					938

Material	Integrated Pest Management (IPM) strategy				
	EIQ	ai	Dose	Applications	Total
Nova 40WP	41.2	0.40	0.3	4	20
Captan 50WP	28.6	0.50	3.0	1	43
Dipel 2X	13.5	0.06	1.5	3	4
Sevin 50WP	22.6	0.50	3.0	1	34
Guthion	43.1	0.35	2.2	2	66
Total environmental impact					167

Material	Organic pest management strategy				
	EIQ	ai	Dose	Applications	Total
Sulfur	45.5	0.90	6	7	1720
Rotenone/pyrethrin	25.5	0.04	12	6	73
Ryania	55.3	0.001	58	2	6
Total environmental impact					1799

[1] active ingredient
Source: Kovach et al. 1992

In this example, a conventional pest management approach to growing Red Delicious apples that does not rely heavily on pest monitoring would result in a total theoretical environmental impact of 938 due to pesticides. An IPM approach that incorporates pest monitoring methods, biological control, and least-toxic pesticides would have an environmental impact of only 182. The organic pest management approach, which uses only naturally occurring pesticides, would have a theoretical environmental impact of 1799 according to the model. The environmental impact of the latter approach is much larger than those of the other strategies because of the larger quantities of sulfur required and the more frequent applications needed to provide the same level of control of apple scab in this variety. Using the EIQ model permits IPM practitioners to rapidly estimate the environmental impact of different pesticides and pest management programs at the planning stage. Thus, growers can incorporate environmental effects along with efficacy and cost into the pesticide decision-making process. IPM research programs can use the EIQ model as another method to measure the environmental impact of different pest management and pesticide programs. As newer biorational pesticides are marketed with lower EIQ values and more emphasis is placed on biologically based IPM practices, the EIQ field use ratings will continue to decrease. They may eventually approach zero, resulting in an agricultural production system that is environmentally neutral or benign with respect to pest control.

Scientific and Management Issues for the Great Lakes (John Hartig)

The Great Lakes are unique among global features, representing approximately one-fifth of the total standing fresh water on the earth's surface. The attitudes about and uses of the Great Lakes have evolved from settlement and initial development to their current state, in which multi-stakeholder partnerships are being formed to address and resolve common problems using locally and regionally designed ecosystem approaches. In general, management of the Great Lakes has evolved through a number of distinct stages:
- settlement and initial development (1600s and 1700s);
- exploitation (clearing of forest land for agriculture in the 1800s);
- reactive management (waterborne disease epidemics in the early 1900s, cultural eutrophication in the 1960s and 1970s);
- proactive management (waste reduction or minimization and risk or hazard assessment to prevent toxic substance problems in the 1980s and 1990s); and
- site-specific ecosystem approaches to management, developed and implemented by multi-stakeholder partnerships (future).

Public awareness and involvement have also evolved from initial, firsthand experiences with problems (e.g., waterborne disease epidemics in the early 1900s) to public participation in development and implementation of comprehensive management plans. This increasing public awareness and stakeholder involvement has led to more effective overall management of the Great Lakes Basin ecosystem.

Environmental issues

Presented below are synopses of a number of the major environmental issues facing the Great Lakes. No priority is intended.

Exotic species

Invasions of exotic species are one of the most pervasive and least-understood anthropogenic perturbations of ecosystems. Since the early 1800s, at least 136 exotic species have successfully become established in the Great Lakes (Mills et al. 1993). Twelve new species have been discovered in the Great Lakes since 1980. Species such as the zebra mussel could have long-term impacts on the structure of pelagic and benthic communities and, consequently, significant adverse economic impacts on the Great Lakes region.

Habitat loss

Loss of habitat is well recognized throughout the Great Lakes Basin ecosystem due to shoreline development, deforestation, erosion and sedimentation, water level fluctuations, navigational dredging, contaminated sediments, and other factors. Although the loss of habitat is well recognized, it has not been quantified basinwide. Selected regional examples of habitat loss include these: wetlands in the St. Clair River–Lake St. Clair–Detroit River system declined from 7274 ha in 1873 to 2022 ha in 1973 (Waybrant and Bryant 1990); 60% of the Great Lakes coastal wetlands in the State of Michigan have been lost due to development (Jaworski and Raphael 1978); and 80% of the wetlands in Hamilton Harbor (western Lake Ontario) have been lost due to development (Canada–Ontario 1988).

Self-sustaining fish stocks

Great Lakes fish stocks have been diminished and altered through human exploitation, habitat degradation, and introduction of exotic species. Much has been done to address, reverse, or compensate for this degradation, but much remains to be done. Over a twenty-year period from 1968 to 1988 approximately 300 million salmon were stocked in the Great Lakes. Great Lakes fishery management agencies have estimated that the angling and commercial fishery was worth approximately two to four billion dollars in total economic activity in 1985, including supporting some 75,000 jobs and providing recreation to over four million anglers. The overall management goal has been to secure fish communities, based on foundations of stable self-sustaining stocks, supplemented by judicious plantings of hatchery-reared fish. These communities are expected to provide an optimum contribution of fish, fishing opportunities, and associated benefits to meet needs identified by society for wholesome food, recreation, employment and income, and a healthy human environment (Great Lakes Fishery Commission 1980).

Eutrophication and food web dynamics

The $9 billion U.S.–Canada phosphorus control program has resulted in improvements in the trophic status of eutrophic areas such as Lake Erie, Lake Ontario, Saginaw Bay, and Green Bay and has contributed to nondegradation of oligotrophic areas (e.g., upper lakes). Concurrently, nitrogen to phosphorus (N:P) ratios have increased. Questions are now being raised about whether or not phosphorus limitation is negatively impacting fish productivity in Lake Ontario. Changes in food web dynamics have also occurred as a result of top-down controls (i.e., fish stocking) and species invasions.

Persistent toxic substances and human/ecosystem health

Considerable concern has been expressed about persistent toxic substances in the Great Lakes because they have become widely dispersed and have bioaccumulated in plants and animals, including humans. Numerous adverse biological impacts have been observed or associated with persistent toxic substances. The weight of scientific evidence indicates that persistent toxic substances are affecting reproductive success of fish-eating birds in the Great Lakes Basin (Giesy et al. 1994). Considerable scientific and societal debate is now occurring over human health consequences of exposure to these contaminants.

Air quality and long-range transport of toxic substances

Toxic substances enter the Great Lakes region as a result of both local activities and long-range atmospheric transport from other regions. The lakes serve as a sink for contaminants deposited on either land or water. Despite reductions in air emissions, there is increasing concern for atmospheric deposition of airborne inorganic and organic pollutants to the Great Lakes. Strachan and Eisenreich (1988) estimated that the atmosphere contributes approximately 90% of the polychlorinated biphenyls (PCBs) to Lake Superior.

Water level fluctuations

In 1985 and 1986, after about two decades of above-average precipitation and below-average evaporation in the Great Lakes Basin, all of the Great Lakes with the exception of Lake Ontario reached their highest levels of this century. Storm activity combined with these high levels to cause extensive flooding and erosion of lake shorelines and severe damage to shoreline property. Millions of dollars of damage resulted. Currently, water levels have decreased toward mean levels. Flooding and erosion caused by wind, wave, ice, and storm action will continue to occur along the shorelines of the Great Lakes, regardless of lake level regulation. No one management measure will address all water level-related problems, nor can management measures be applied in specific instances without regard for measures taken in other areas or without regard to the varied interests affected.

Chapter 3: Case studies

Global phenomena

Increasingly, environmental studies are concerned with the impacts of human activity on a global scale. Some include predicted global climate changes associated with the "greenhouse effect" and the increase in ultraviolet radiation reaching the surface of the earth as a result of the destruction of the ozone layer. In general, models used to forecast these events are relatively imprecise and fairly coarse in their spatial and temporal precision. Accurate forecasts of regional climate change in response to global atmospheric changes are not currently feasible (Houghton 1995), but any significant change in temperature or annual rainfall related to global warming would have major ecological and socioeconomic impacts on the Great Lakes region.

Management processes

Presented below are synopses of selected current management processes for addressing some of the major environmental issues facing the Great Lakes. No priority is intended.

Virtual elimination of persistent toxic substances

The International Joint Commission (Strachan and Eisenreich 1988) created a Virtual Elimination Task Force to address and advise on the Great Lakes Water Quality Agreement's commitment to virtually eliminate the inputs of persistent toxic substances. A broad-based task force of U.S. and Canadian persons representing numerous sectors was formed. Terms of reference were issued and the task force adopted a vision (i.e., women can eat Great Lakes fish without affecting the development of their babies; wildlife that eat Great Lakes fish and other aquatic life thrive in the basin; and people can eat Great Lakes fish without increasing their risk of getting cancer). Using this multi-stakeholder process, a virtual elimination strategy is being developed that addresses pollution prevention, banning of certain chemicals, and phaseout procedures. Strengths of this International Joint Commission process include

- common fact-finding by groups of experts representing numerous sectors in both the U.S. and Canada, serving in their personal and professional capacities; and
- public participation to obtain valuable input and build confidence and support for decisions.

Sustainability of Intensively Managed Populations in Lake Ecosystems

Sustainability of Intensively Managed Populations in Lake Ecosystems (SIMPLE) is a study being undertaken as a project of the Great Lakes Fishery Commission's Board of Technical Experts to address the long-term sustainability of hatchery-dependent salmonid fisheries on the Great Lakes. Among the issues addressed are these:

- limits to salmonid stocking,
- the changing salmonid community, and

- sustaining the miracle of economic and social benefits derived from the salmonid fishery.

SIMPLE is a process of workshops and modeling exercises aimed at assisting fishery managers to find more sustainable management practices. Technical experts, fishery managers, and other interested stakeholders use modeling outputs to identify a set of plausible alternative futures based on realistic sets of management decisions and their predicted consequences.

Remedial Action Plans

Currently, 43 areas of concern have been identified in the Great Lakes where beneficial uses are impaired. Broad-based institutional structures are being used to help build the capacity of governments to develop and implement comprehensive and systematic remedial action plans (RAPs) (Hartig and Zarull 1992). These multi-stakeholder groups are reaching agreement at key points in the decision-making process (e.g., problem definition, selection of remedial and preventive actions). Progress is sustained by an iterative process of establishing short-term priorities and celebrating milestones. The result is action planning within a strategic framework.

Ontario Roundtables on Environment and the Economy

Ontario has established a provincially led process of linking environment and economy in order to achieve sustainable development. Essential characteristics of such multi-stakeholder roundtable processes include being inclusive, emphasizing cooperative learning, sharing decision-making power, having independent facilitation, establishing local ownership, providing a forum for mutual understanding and new perspectives, establishing trust, encouraging a positive reinforcement system, committing to mutual accountability, and assessing and celebrating progress. Scientists play critical roles in defining environmental problems, establishing cause-and-effect linkages, communicating remedial and preventive options, providing a "reality check," and helping make midcourse corrections.

Great Lakes Initiative

The Great Lakes Initiative is being heralded as a landmark effort in the control of toxic substances in the Great Lakes Basin (Whitaker 1993). Regulatory guidance is being developed for protection of aquatic life, human health, and wildlife. It attempts to achieve a level playing field. Each of the Great Lakes states must amend its water quality standards to be consistent with this "mandatory guidance" within two years following final publication, or USEPA will promulgate the rule itself. In order to be consistent, each state's plan must be at least as stringent as the Great Lakes Water Quality Guidance for every part of the guidance. Procedures governing mixing zones and waste load allocations emphasize the total maximum daily load approach. Strengths of the Great Lakes Initiative process include technical content, consistent approach to reducing risk, inclusivity, and public participation. Challenges include streamlining the process and reaching agreement on costs and benefits.

Conclusions

Numerous individuals argue that SEM, an ecosystem approach to management, and sustainable development are wonderful in theory and difficult in practice. Greater emphasis must be placed on understanding ecosystem structure and function. Greater emphasis is being placed in Great Lakes management on "anticipate and prevent" and ecosystem-based management initiatives. In the real world of resource management, there is a considerable lack of proof of causality and substantial uncertainty. Therefore, decisions are being made based on the best available information and knowledge.

A number of new management initiatives have been undertaken in the Great Lakes Basin that attempt to balance idealism with realism. Common characteristics include the following:
- establishment of multi-stakeholder institutional frameworks to address issues,
- quantification of problems and establishment of priorities using risk assessment,
- agreement on a long-term vision and short-term measurable goals and objectives,
- evaluation of management alternatives using modeling or other techniques,
- implementation of high-priority actions,
- use of a two- to five-year iterative process for reassessing problems and priorities and for implementation of new high-priority actions,
- measurement of progress toward the long-term vision,
- measurement of stakeholder satisfaction, and
- celebration of successes on a routine basis.

Meeting the challenge of SEM will require greater institutional cooperation, coordination, and integration of research, planning, and program activities and functions. Multi-stakeholder, multi-sectoral, institutional frameworks represent a forum for cooperative learning to generate a common understanding of problems and to build consensus for action. Such institutional frameworks must be empowered to pursue a vision and must not be rule-driven. Empowerment of institutional frameworks would be demonstrated by these characteristics:
- a focus on watersheds or bioregions,
- an inclusive and shared decision-making process,
- clear responsibility and sufficient authority to identify and pursue a road map to problem resolution and goal attainment,
- a strong commitment to broad-based education and public outreach, and
- an open and iterative process that strives for continuous improvement.

Long-term thinking, scientific information, and sociopolitical receptiveness (social acceptability and political feasibility) are fundamental prerequisites to SEM. Given the environmental management time frames and scientific uncertainties discussed above, sociopolitical uncertainty may slow the onset of SEM.

It is important to note that at all levels, general public support is absolutely essential to the success of ecosystem management efforts. A major source of uncertainty is the extent to which public support can be rallied around preventive rather than curative tasks. Active public education and outreach are therefore of paramount concern in successful landscape and watershed management efforts.

Sustainability-based Forest Management in Michigan's Upper Peninsula (Dean Premo)

This case study is drawn from a cooperative effort between industrial forest managers at Mead Corporation and ecologists at White Water Associates, Inc. It describes the process of sustainable forest management by defining the challenges, outlining the approach adopted for meeting the challenges, and presenting some on-the-ground examples.

Defining the challenges

Sustainability in the context of managed forests means maintaining long-term productive potential, implying that all ecosystem components must be maintained (Mladenoff and Pastor 1993). This represents a daunting charge to a resource manager. Traditionally, foresters were responsible for managing forests for wood production, but today they are confronted with a plethora of new challenges, including rare and endangered species, biodiversity, forest fragmentation, neotropical migrant birds, game and nongame wildlife, water quality, wetlands, and ecosystem management (Rogers and Premo 1992). Shouldering these new responsibilities, forest managers appeal to scientists for practical sustainability-based forest management techniques. In responding to this appeal, we are continually reminded that a sizable gap exists between ecological science and practical forest management. Scientists (in this case, biologists) have done well at accumulating information but have done poorly at conveying it to managers.

Meeting the challenges

Mead Corporation owns and manages 700,000 acres in the Upper Peninsula of Michigan, representing about 7% of the surface area. Approximately 40 Mead staff are involved with the various aspects of this management. In 1989, Mead approached White Water to help implement its stewardship policy that balanced commodity production with protection of other forest values. Mead expressed a need for ecological information and ways of incorporating ecology into their forest management. From this initial contact has developed the ongoing Total Ecosystem Management Strategies (TEMS) program. This interactive research and education endeavor is in its fourth year.

In designing TEMS, White Water considered an interactive suite of four elements. The starting point was an examination of the underlying philosophy of the landholder. Obviously, land management philosophy varies among landholders such as the U.S. Forest Service, the Nature Conservancy, or Mead Corporation. Second, White Water investigated policies and procedures that have direct influence over Mead's forest management activities to determine how these policies might help or hinder SEM and how they could be modified. Third, White Water studied the forest managers who effect the philosophy and policies at Mead, including their backgrounds, motivations, incentives, and continuing education needs. Finally, the landscape being managed was investigated to determine its expanse, composition, history, and ecology. These four elements provide a basis for integrating ecology, landscapes, and people in a dynamic, adaptive process of forest management.

Biodiversity is the focus for TEMS. Biodiversity is both measurable and associated with ecosystem health, making it a logical integrator of ecology and SEM. Such integration has not traditionally been included as part of a forestry training curriculum, even though a forester's decisions affect all components of an ecosystem. By the same token, ecologists are rarely trained in the field of forestry and are unaware of management options available to foresters (Rogers and Premo 1992).

Forest management is a continuous and ongoing process that cannot be suspended as society waits for more studies and better information. Quite simply, forest management, both good and bad, is done by foresters who will apply new strategies as they are developed. Foresters, therefore, have to be a primary focus for implementing SEM. For this reason, TEMS has taken a bottom-up approach. Foresters are not, however, working in isolation; decision-makers above them affect forest management. As Walters (1986) pointed out, these decision-makers are real people guided by diverse motives not necessarily in accord with simple objectives. These higher-level decision-makers (administrators, accountants, corporate vice presidents) are also an important educational audience. Finally, there are woods workers who carry out the forester's plans. The road builders, equipment operators, and loggers are also in need of ecologically based information pertaining to the work they conduct. In the TEMS program, these diverse audiences are addressed with a variety of techniques.

Information transfer is conducted through personal contact and written communications. One of the most constructive and dynamic educational tools is direct interaction between Mead forest managers and White Water ecologists in the form of day-long field trips, interactive workshops, one-on-one discussions, and participation by foresters in White Water research projects. Each contact is treated as a two-way education opportunity. For example, each year White Water ecologists and Mead forest managers take several landscape-level field trips. Each field trip is preceded by a review of Mead's geographical information system (GIS) maps and discussions with the area forester in charge of a several-thousand-acre landscape. An itinerary is established, with planned stops that help illustrate specific landscape challenges faced by the area forester or that allow White Water scientists to discuss topics pertaining to the ecology of the given landscape (including dispersal corridors, rare plant habitat, wetland issues, and herbivore damage). Each field trip is followed by a written report from White Water and a five- to ten-year plan prepared by the area forester that incorporates concepts and techniques discussed in the field.

Written materials are also important mechanisms for information transfer. The TEMS program has produced three types of documents: a study guide, a periodical publication called "Strategies," and research reports (White Water 1992a, 1992b, 1993). All are written in a style that conveys technical information in a common-sense format and that is usable by forest managers.

A second important thrust of the TEMS research–education approach is applied research, including both gathering existing ecological information and conducting field research. As Walters (1986) noted, management activities themselves are primary tools for experimentation. An abundance of research opportunities is afforded by the whole-system manipulations performed by forest managers. White Water has conducted several research projects that have immediate and direct applications by forest managers. Examples include the following:

- bird community composition in forest stands of diverse management history,
- bird and mammal use in selectively harvested stands,
- winter-season mammal use of riparian areas,
- breeding bird density in streamside buffer zones of varying widths,
- water quality and stream biodiversity relative to buffer zone width, and
- landscape values of vernal ponds.

In addition to providing important applied information, research projects are valuable educational resources that allow ecologists to bring further appreciation of ecology to foresters.

On-the ground examples

Forests and nesting guilds

In winter 1990, White Water ecologists, using snowmobiles and snowshoes, established 24 bird sampling plots in diverse cover types of a 20,000-acre landscape. In spring 1991, standard techniques were used to systematically census birds on these plots; 93 species were recorded. These data were used later to examine patterns of biodiversity. In the later part of the field season, White Water took small groups of Mead personnel to selected sample plots for demonstrations of sampling methods. For many foresters, the diversity of birds present on their very familiar forested landscape became a real, aesthetic experience for the first time. The large number of ground- and shrub-nesting species was a revelation to the foresters. They began to question the traditional practice of removing understory balsam fir (an alleged competitor) during selective hardwood harvest. Through feeding guild analysis of the bird data, there was further recognition that retaining maximum diversity of insectivorous birds also maintained a great potential for insect control in the forest. Mead foresters now plan for better understory structure.

Riparian management

Areas along streams and rivers are of enormous concern to forest managers. The typical focus for this concern, however, has been buffer zones established for reasons of water quality and aesthetics. In TEMS, White Water has broadened this perspective to consider the terrestrial values imparted by riparian forests and vegetation. Researchers have investigated questions regarding the amount of streamside forest cover necessary to maintain ecosystem values.

In one study, White Water (1992b) used winter tracks to assess mammal species diversity and use in the main forest types occurring along the West Fence River. Employing data on track density and numbers of species, ecologists sought to determine natural zones of winter use by mammals relative to distance from the stream. From this study, White Water tentatively recommended that buffer zones of 100 feet would optimize benefits as mammal travel corridors in riparian hardwood stands. In conifer stands, mammal use peaked at 400 feet from the river, leading to a recommendation that wider buffer zones were desirable for this cover type. These wider nodes could be connected by narrower corridors of 100 feet, thus establishing a buffer zone resembling a beaded necklace. This pattern is similar (although on a smaller spatial scale) to the nodes, networks, and multiple-use modules of Noss and Harris (1986).

In a related study, breeding bird variety and density were assessed along the forested river corridor of the Fence River. The objective of this work was to determine whether a relationship existed between bird community structure and proximity to the river. The study was also designed to allow examination of the riparian bird community following clear-cut harvesting that left buffer zones of 50-, 150-, and 250-foot widths.

Fence ford

As the TEMS program has continued, White Water has witnessed Mead foresters expanding their professional horizons. At the Fence Forest landscape, a mixed stand of hardwoods and conifers was scheduled for clear-cut and later conversion to red pine. Integrating information from TEMS research and education, the area forester exercised sound ecological judgment regarding a vernal pond and habitat islands as he planned for the clear-cut.

In the spring, inconspicuous vernal ponds form ephemeral oases of aquatic and terrestrial life. Ecologists suspect that their presence contributes greatly to the overall forest ecosystem by providing habitat for a variety of aquatic invertebrates, a breeding and feeding site for many amphibians and reptiles, an attractive feeding and resting spot for songbirds, a source of food and water for many mammals, and a unique microhabitat for plants. Despite this importance, most forest management plans do not afford vernal ponds any special management attention. Because they are often dry, open spots during logging operations, these ponds become places to deck logs and pile slash. As an experiment, the area forester decided to leave a 50-foot buffer zone of trees and natural vegetation around the pond and a 500-foot forested corridor between the pond and intact hardwood forest. Pragmatic factors supported this decision. The value of the timber in the buffer strip and corridor was fairly low, and wet soils would have made harvesting difficult and soil damage likely.

Reflecting on TEMS research and discussions with White Water ecologists, the area forester marked some patches of dense young spruce, fir, and aspen as areas to leave uncut. Earlier that year, research on the nearby West Fence River demonstrated heavy mammal use in small habitat islands that inadvertently had been left standing in clear-cut areas. This practice has since been replicated on other Mead lands.

Mulligan Creek Coalition

As the forester's perspective shifts from managing forest-stand yield to managing ecosystem processes, the importance of considering large landscapes emerges. This evolution reveals the necessity of involving several or many owners in cooperative management. The Mulligan Creek Coalition is an example, spawned from the TEMS program, of the initial steps of a landscape management partnership based on the initiative of three corporate landowners: Mead Corporation, Champion International, and Longyear Realty.

Mulligan Creek runs approximately 20 miles through north Marquette County, Michigan. A large component of the Mulligan Creek area (approximately 18,000 acres) is owned by Mead, Champion, and Longyear. The Mulligan Creek area provides significant, large-scale landscape linkages with respect to plant and animal migration and dispersal in Michigan's Upper Peninsula and the larger Lake Superior Watershed. It is an area where the gray wolf (Michigan Endangered) and possibly the Canadian lynx (Michigan Endangered) reside and travel. The Mulligan Creek area has not been intensively studied by scientists, yet a few casual observations indicate that the area harbors an abundance of rare plant species and many specialized habitats (Premo 1992).

Mead forest managers asked White Water to prepare a conceptual plan for maintaining the important ecological features of the Mulligan Creek landscape while managing their forests primarily for high-quality hardwood products. Based on White Water's recommendations, a Mulligan Creek Riparian Management Area (RMA) was established and subdivided into three functional zones: interior, forest, and matrix. The interior includes Mulligan Creek, its tributaries, and the gorge (all areas of high potential for rare plants). The forest includes corporate landholdings that are contiguous with the interior. The matrix comprises all of the lands surrounding the interior and the forest.

White Water recommended that the interior not be entered for timber management or harvest and that natural features within the interior should be characterized, mapped, and monitored. The forest is the portion of the Mulligan Creek RMA where forest management and harvest can be conducted. The forest provides an important protective function by attenuating human impacts on the interior. White Water recommended that management and timber-harvesting activities in the forest be designed to progressively minimize ecological impacts and human intrusions, moving from the outer border of the forest toward the interior. Only limited opportunity exists to directly manage the matrix because it is outside of the coalition's ownership, but its composition and human uses are important to monitor (Premo 1992).

Mead, Champion, and Longyear now maintain an integrated GIS database on the watershed. They are cooperating on road construction, maintenance, and closure. Approximately 97% of the landscape will be managed by selective harvest to maintain a closed-canopy northern hardwood forest. Three percent of the area (about 500 acres of aspen and swamp conifer) will be managed by even-aged management over the next ten years. Wide, forested buffer zones (interior) have been established around Mulligan Creek and its tributaries and around inland lakes in the watershed (White Water 1993).

In 1992, Michigan's governor recommended that the Michigan Department of Natural Resources (MDNR) cooperate with the Mulligan Creek project. Although State of Michigan land ownership in the watershed is quite small, the MDNR has jurisdiction over many of the resources present (e.g., rare plants and animals, fish, and game). The three corporate participants are also encouraging small-acreage owners in the watershed to become part of the coalition. The Mulligan Creek coalition advances landscape-level management in several important ways:

- It contributes to developing practical management paradigms and techniques.
- It provides an outdoor classroom for educating many audiences.
- It presents a pilot project for discovering and solving the challenges of multi-owner cooperative management efforts.
- It provides a palette on which resource managers with diverse interests can blend management perspectives.
- It represents a demonstration project for the public to view and consider.
- It fosters the formation of other landscape coalitions crucial to sustainability-based forest management.

Product Development (Sven-Olof Ryding)

Solving product-related environmental issues should lie at the heart of future industrial activities. The only way to achieve such an objective is to work in a preventive and cautionary way, based on a pragmatic but holistic approach, from problems to strategy. The only known and experienced concept to deal with all these new, complex issues seems to be a life-cycle approach. LCAs have been widely recognized as an environmental management tool for the future (Fava et al. 1991). Although LCAs promise to be a valuable tool for programs on cleaner production strategies and design for the environment, the concept is relatively new and will require future refinement to be accepted as a general tool for SEM.

To meet sustainability-based demands on future product development and to guide industry to adopt LCA-related thinking into commerce, the Federation of Swedish Industries initiated and now coordinates a joint project activity (Steen and Ryding 1992), together with 15 major companies, the Chalmers University of Technology, and the Swedish Environmental Research Institute to develop a user-friendly in-company tool for Product Life-Cycle Assessments (PLCA).

The system, called the "Environmental Priority Strategies (EPS) for product design system," is intended to help product designers, purchasers, and decision-makers incorporate the environmental impact of processes and products at an early stage of planning and product development. The system follows the overall framework for LCAs suggested by SETAC (Fava et al. 1993). However, the EPS system contains supplementary components especially designed to fulfill sustainability-based criteria, including these:

- a holistic, environmental impact assessment approach that weighs the environmental effects caused by use of natural resources together with those traditionally caused by pollutant emissions;
- a recycling concept that considers a quantitative and qualitative environmental understanding and accounting of various flows of virgin and reused materials; and
- a sensitivity and error analysis approach to elucidate the effects of uncertainty in underlying assumptions and input data for decreasing the risk of making false conclusions about the relative harmfulness of alternatives in product development.

The EPS system is designed for industrial use by purchasers and manufacturers through the successive definition and buildup of environmental load indices for construction materials and manufacturing processes.

Technical framework

The main idea of the EPS system is to make environmental loads and environmental impacts of products visible through a transparent calculation procedure that synthesizes and integrates environmental concerns. Consecutive LCA components — initiation, inventory analysis, impact assessment, and improvement assessment — are clearly displayed, enabling users to choose any desired level of complexity. The EPS system provides a method for enumerating and assessing ecological and health consequences of human activities such as pollutant emissions, energy use, raw materials extraction, and land use before aggregation and final evaluation. Both basic assessment of values of environmental qualities and changes in these values from human activities are estimated.

The overall purposes of the EPS system are these:
- to describe the environmental impacts of the consumption of energy and raw materials and of pollutant emissions during the different phases of a product's life cycle,
- to systematically provide information useful for an aggregated environmental impact assessment of products from cradle to grave, and
- to evaluate the environmental consequences of alternative processes and construction concepts, in relative terms, to enable comparisons between different process approaches and product designs.

The EPS is a "bookkeeping system" of environmental loads and environmental impacts. The principal tools are the definitions of so-called environmental load indices for use of natural resources and raw materials and for various pollutant emissions to air, soil, and water. The "environmental load index" represents a valuation and weighing of the importance attached to the use of a selected resource or to the emission of a certain pollutant. Impacts to be valued are described in terms that are compatible with both the knowledge structure of environmental sciences and existing valuation systems. The following stepwise procedure is used to place values on environmental impacts:

1) Define "unit effects" for the five safeguard subjects: biodiversity, human health, production, resources, and aesthetic values.
2) Value the unit effects according to a "willingness to pay" concept.
3) Set a value to changes in the unit effects.
4) Estimate the contribution of the resource use, pollutant emission, or human activity being evaluated to this value of change in the unit effects.

Impacts on the five safeguard subjects are valued on a relative scale in environmental load units (ELU) according to the willingness to pay for avoiding negative effects on the safeguard subjects. One ELU is meant to correspond roughly to one European currency unit (ECU) in the OECD countries.

The background information originates from an LCA-based inventory of the materials and processes under study. Employing the indices for use of natural resource and raw materials, and for emission to air, soil, and water, the indices are combined with inventory data on 1) consumption of energy and raw materials and 2) pollutant emissions for the product life cycle. The resulting values form a database of environmental load indices for construction materials, substances, elements, or semi-manufactured products. The dimension obtained is usually ELU/kg, but other units may be used, e.g., ELU/m^2 or ELU/piece.

The database of environmental load indices for construction materials and manufacturing processes forms the vehicle for life-cycle analysis in the EPS system. The user of the system is provided with the basic input to the simple formula used for calculating an environmental load value:

Environmental load index × Quantity = Environmental load value.

The environmental load value for an activity or product life cycle is expressed in ELU, a dimensionless number to be used for guidance on the environmental aspects of a process or a product. The lower the environmental load, the lower the overall environmental impact. Any such number may be broken down into its principal components for further analysis; the user can choose the desired level of complexity. In the case of recovery or recycling of material, or cogeneration of energy by incineration of wastes, a value equivalent to the environmental benefit of recycling or cogeneration is subtracted at the end of the calculations.

The uncertainties inherent in the data input and the criteria for valuation are dealt with in the EPS system by assigning standard deviations to these data and criteria. Sensitivity and error analyses provide insights into the influence of this uncertainty on the comparisons and decisions at hand, and they permit calculation of the probability that one process or product alternative is more environmentally benign than a competitor process or product.

The EPS system is intended to be a universal tool for performing environmental impact assessments for various types of products and human activities. The approach used to develop the EPS permits calculation of environmental loadings for a variety of products made of complex materials and for construction materials used by a variety of industries. For example, materials involved in construction of a sofa may include timber, textiles (for the cloth), plastic, and metal. Environmental load indices for all of these materials are included in the EPS database.

A universal system for assessing the environmental impact of products facilitates communication between various representatives in the business sector, including purchaser, product developer, designers, marketing representatives, and board members. It also should raise the quality of the dialogue and understanding between the business community and the public. Current Product Ecology Project activities include the following:

- development of a code system for classification of environmental load indices,
- further refinement of the EPS Enviro-Accounting Method (Steen and Ryding 1992),
- definition of the first complete version of environmental load indices,
- clarification of terminology and definitions,
- development of a recycling model as part of the calculation procedure,
- fine-tuning and combining the two data software tools into one LCA inventory and evaluation tool,
- development of a proposal for future administration of the EPS system,
- production of information materials and standard educational packages, and
- preparation of a handbook on environmentally sound product development.

The Everglades (Larry Harris)

The area of Earth presently known as the Everglades consists of a complex, interactive set of component subsystems that formerly did, and perhaps in the future will, function as an integrated system, more specifically a spatially heterogeneous ecosystem. This articulated set of subsystems contained the following notable features:
- one of the single greatest wetlands on earth, certainly the largest in the lower 48 United States;
- Lake Okeechobee, the second largest lake in the lower 48 states;
- a collection of lakes in central Florida (near Orlando) known as the "Upper Chain of Lakes";
- the Kissimmee River, a 140-km, highly meandered drainage from the upper chain of lakes southward to Lake Okeechobee;
- a largely forested area of southwest Florida referred to as the "Big Cypress"; and
- a large estuarine area, Florida Bay, immediately south of the Florida peninsula.

The Everglades region is the only part of the continental United States that falls in the subtropical life-zone. Because it was so profusely colonized by flora and fauna of West Indian (Antillean) origin, it supports a tree species diversity greater than all of Europe and about as great as all of the rest of North America. The Everglades ecosystem was and is dominated by primitive taxa that predate the demise of dinosaurs. These include the water lily and water lotus, cypress and magnolia, alligator and crocodile, bowfin and gar, the Snake bird or Anhinga, sharks, and marine turtles. These species have survived all of the extinction spasms that occurred during the last 70 million years.

In 1850, about two-thirds of the area south of Orlando was contained within the Kissimmee–Okeechobee–Everglades/Big Cypress–Florida Bay system. That year, the U.S. Congress passed the Swamp and Overflowed Lands Law of 1850, which conveyed ownership of this land to the state of Florida on the specific condition that it would be "reclaimed." In 1850, reclamation was understood to mean drainage for the purpose of productive human enterprises such as agriculture. Most of the area was sold to the private sector for less than $1.00 per acre, and both the vanities and fortunes of several prominent Americans were crushed as they attempted to "reclaim" this massive landscape.

Understanding how the various subsystems interact in time and space is critical to any appreciation of the complexity of the system's current dysfunction. For example, the American alligator reaches sizes that are best measured in meters of lengths and hundreds of kilograms. As a means of survival during the seasonal dry periods, these animals either seek or create water-holding depressions that are referred to as "alligator holes." Historically, the rapid drydown of the Everglades subsystem caused the concentration of millions of fish in localized areas such as alligator holes and triggered the breeding cycle of a now-endangered species, the wood stork. If the drydown and concentration of fish does not occur quite rapidly, the wood storks either fail to initiate breeding or fail in their attempts. Over a dozen other species of such wading birds as herons, egrets, ibis, and spoonbills also forage in these highly localized depressions that contain both water and fish throughout the dry season.

Also dispersed across the Everglades are a myriad of small, slightly elevated islands dominated by trees. During flood stage, these tree islands serve as critical refugia for terrestrial species that would otherwise succumb due to the flooded landscape. Hundreds of thousands of colonial-nesting wading birds, for which the Everglades is highly renowned, also roost and nest in these tree islands. One million wading birds transfer approximately 10,000 tons of nutrients from their aquatic foraging areas to their roosting sites during the course of a decade. Even at the scale of a few hundred km^2, subsystem components interact in time and space to allow the functioning of the larger system.

At a larger spatial scale, i.e., the 250-km north-to-south drainage, subsystem interactions are equally necessary. Precipitation in any part of the system is important to that particular area, but the entire watershed must be evaluated in order to understand both the functioning and the present plight of the Everglades complex. Historically, when heavy precipitation events occurred in the Orlando area, the combined Upper Chain of Lakes and the highly meandered Kissimmee River basin (as wide as 15 km across) stored millions of cubic meters of water that was slowly released southward to Lake Okeechobee, Everglades–Big Cypress, and finally Florida Bay. Full execution of such a phenomenon spanned several months and laterally flooded thousands of acres. The front edge of the shallow floodwaters constituted a migrating ecotone between aquatic and terrestrial subsystems. Like a python digesting a pig, a slow but inexorable wave of nutriment moved through the system. Large, long-legged wading birds such as wood storks are near the top of the food chain, and they easily fly 50 km or more to access an appropriate foraging area. A bulge of flood water attempting to move 200 km through a landscape of dense marsh and forest requires months and constitutes a massive time–space interaction that was necessary for the diverse and abundant fauna for which the Everglades region is renowned.

Today, the Everglades regional system is ecologically dysfunctional. Much of that present dysfunction derives from disarticulation of the previously connected subsystems, and some of it also derives from massive land use and management changes within three or more of the subsystems. Finally, some of the current dysfunction derives from totally exogenous forces such as mercury loading and even higher-scale impacts deriving from global climate change and sea level rise.

The following items will indicate the degree of "reclamation" that has occurred, mostly in the last 50 years, and some of the principal obstacles that must be overcome to permit sustainable management of the Everglades:

- The Upper Chain of Lakes have had their water levels lowered and are no longer able to support the same volume or quality of water as before.

- The highly meandered and seasonally flooded Kissimmee River basin has been channelized, resulting in a rapid flow of water southward and almost total elimination of a marshland system that functioned as a nonstructural hydrology and water quality manager.
- Over 2250 km of canals with attendant levees ensure expeditious purging of flood waters to the sea, thereby disrupting natural hydroperiods and flow regimes.
- Over 100 flood control structures, when coupled with political forces at work in distant lobbies, guarantee water allocations based only partially on ecology and environment.
- The south Florida human population now exceeds 3 million and demands fresh water not only for consumption but also for the purpose of preventing saltwater intrusion.
- The nearly 50,000-acre Everglades Agriculture Area requires fresh water for agricultural production, while discharging nutrient-rich effluent that threatens the integrity if not the survival of other subsystems such as Everglades National Park.

Recreating a functional and sustainable Everglades system will require at least three different levels of attack. At the highest level, the issues of climate change, rising sea level, the need for entire life zones to migrate spatially, and the need for Florida Bay to be considered as more than a receiving area for Everglades runoff must be recognized. Although climate change and sea level rise occur almost imperceptibly, disturbance events such as hurricanes may increase. It might be anticipated that massive effects will be pursuant to storms of the future, even though the same magnitude of storm might have had much less impact in the past.

At a somewhat lower scale of resolution, the issue of mercury loading from an overarching airshed must be dealt with. It is also at this scale that the rearticulation of the various subsystems must be achieved. At least two north–south corridors that reconnect the lower Everglades with the Kissimmee and/or the St. Johns River basin must be created to restore the natural system function that Lake Okeechobee and its surrounding littoral communities formerly performed. These articulating corridors not only will facilitate the migration of species and life zones endangered by climate and sea level change but also will serve more immediate needs of water delivery and Florida panther survival. This is also the relevant scale for restoration of the regional hydroperiod, for not only the temporal oscillation at any single point but also the spatial movement across any given point in space.

Although they occur at a smaller spatial scale, several local issues have such intense impact as to demand immediate redress. Examples of these include 1) phosphorus pollution derived from the Everglades agricultural activities and abatement and 2) ultimate elimination of many exotic species. In total, large-scale, longer-term restoration (e.g., rearticulated subsystems) must be the backdrop against which medium-scale, midterm issues such as hydroperiod are evaluated and must also be the context within which the smaller-scale, but intensely pressing, local problems must be addressed.

Chapter 4

Steps toward Sustainable Environmental Management

Certain aspects of ecosystem management are best suited to implementation on a national or international basis for environmental, political, or economic reasons. Indeed, some measures can be implemented effectively only at the national or international level. For example, management of global climate change may be a priority concern for the protection of the Everglades ecosystem but is virtually impossible to effect at the ecosystem level. Human population growth is another environmental management problem that regions depend upon nations to resolve. Ultimately, in order to manage landscapes and watersheds sustainably, there is a need to manage the size of the human population that relies on the resource. The most effective way to do this, short of coercion, is to improve education. Therefore, a general goal for SEM should be to improve educational programs. Furthermore, in order to ensure the success of any SEM program, an educational campaign on the meaning and importance of the concept of sustainable development will be essential.

Economic factors also introduce uncertainties that are often best resolved at the national and international levels. For example, adversely affected interests may work in opposition to ecosystem management, supporting their positions with the existence of scientific uncertainties and effectively intimidating lower jurisdictional units. National or international commitments to ecosystem management measures are important shelters from political fallout for state and local implementors of ecosystem management strategies. In addition, different ecosystems may have different sensitivities to loadings of a given pollutant, but nonuniform pollutant standards can create an uneven disadvantage for various stakeholders, with resulting inequitable economic outcomes for industry and host regions. National and international agreements on environmental quality criteria, substance bans, etc., may be necessary to prevent geographic market distortions.

Dissemination of Sustainable Technologies

A significant step toward global sustainability of economic and environmental conditions could be taken by mobilizing experts drawn from a variety of disciplines and organized into cross-functional teams to address specific needs such as industrial pollution control, energy efficiency, and forestry management in developing countries. Individual experts or teams of experts could provide many valuable services, as the following examples illustrate:

- Conduct analyses of production sectors with specific attention to changing current practices to reduce environmental or ecological impacts.
- Provide technical assistance for specific industries, in LDCs, or domestically within the U.S, to examine opportunities for relatively low-cost, high-benefit environmental and energy improvements.
- Assist in building an environmental management system adapted to a specific industry's or country's level of need.
- Help build a capability for continuous assessment of environmental and energy needs.
- Provide assistance and advice on technologies and other options to remediate ecosystems experiencing high levels of stress.
- Provide a mechanism for identifying market opportunities for U.S. technologies and products.

The potential recipient (i.e., host) countries vary greatly in their stages of development as well as their affluence; the information and technologies that could be transferred vary accordingly. For example, some nations in the Middle East are just now starting to implement environmental management programs and are in need of technical assistance and advice. In this instance, financial constraints may not be as limiting as they would be in some Southeast Asian or African countries. Likewise, some nations basically have no environmental programs, and the information they need at the start is very rudimentary. Countries or regions in the United States chosen for involvement in the program should be a mix of those where the greatest progress can be achieved with the least investment of time and effort (i.e., picking the "low-hanging fruit") and those in the greatest need.

One of the hallmarks of sustainable resource management, productive processes, and, indeed, all economic activity must be their participatory nature. Provision of assistance to any host country can be successful only with the host's full commitment and participation. Identification of relevant issues would occur through a participatory process featuring cooperation among scientific experts, public officials, policy-makers, and environmental managers, and private sector representatives. One important attribute of this needs assessment process is that it should involve multinational representation to the greatest possible extent, increasing the probability of creating cooperative international solutions to transboundary issues.

Needs assessments for candidate countries should consider various criteria for judging the likely success of environmental technology transfer. These criteria could include the following:
- presence of highly stressed, unique, or economically important ecosystems;
- host country receptivity;
- existing infrastructure;
- industry presence; and
- understanding of priorities and needs of the host country.

Scientific knowledge and tools developed in the United States or other developed countries must be applied in a fashion consistent with recipient country's technical capabilities. Opportunities for research in host countries should be sought on the basis of specific environmental and energy problems for which the results could be generalized and applied to other potential host countries. For example, the toxicological and chemodynamical data that support regulatory standards developed in the U.S. and the European Community may not be applicable to the tropics. Appropriate studies, conducted in tropical countries possessing sufficient scientific infrastructure to support a cooperative research program, might provide a scientific foundation for tropics-specific regulatory standards.

Organized, technology transfer activity of this kind cannot be an ad hoc activity. Active support from government and/or private industry would be required to establish the necessary multidisciplinary teams. A "Sustainable Development Corps," modeled on the Peace Corps would be one possible model for such an effort, although not the only model. Technical experts could be recruited from SETAC, ESA, other professional societies, and other participating organizations. In addition to active employees of universities, corporations, and nongovernmental organizations, individuals could be recruited from the ranks of retired personnel. Teams of experts could be organized to spend varying periods of time in specific countries or regions of the U.S., lasting from weeks to months. Through these activities, it is hoped that leaders and technical personnel in host countries and the U.S. would understand that greater linkage can exist among energy efficiency, environmental improvements, and economic progress.

Strategy for Sustainable Management of Regions

Sustainable management of regions and watersheds presents a quandary. Because of the large scale of regions and of the environmental problems associated with them, sustainable management must be directed by an entity with a large purview. However, long experience shows us that effective and efficient solutions to environmental problems are best developed by involving local stakeholders who understand the local resources and users of the resources. Figure 3 outlines a general strategy for sustainable management of regional resources that attempts to reconcile these observations.

Broad environmental goals such as attainment of fishable and swimmable waters and preservation of species diversity are set by national and state governments. Some of these goals can be achieved only by national entities, so their agencies must implement them. However, many environmental goals are best accomplished at the regional or watershed scale. Such goals are delegated to the appropriate regions. Because authority is vested in governments, this means that the governmental bodies (states, counties, cities, etc.) within the region are responsible for accomplishing the goal. In some cases, they will be able to implement actions to accomplish the goal. However, in many if not most cases, the needed actions will not be clear to those governmental bodies because they involve

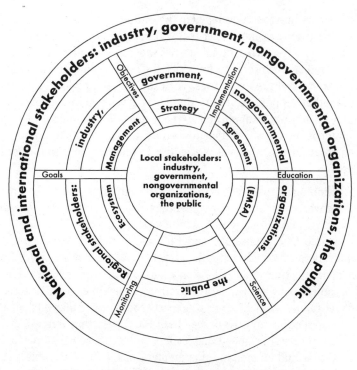

Figure 3 Strategy for sustainable management of regional resources

the region as a whole or because the management actions required are complex. Therefore, these political units must develop a joint approach to achieving the goal. Such an approach may be termed an "Ecosystem Management Strategy Agreement" (EMSA).

An EMSA is arrived at by relevant governments for the purposes of managing human activities in that regional ecosystem and accomplishing particular goals. This could involve authorizing and chartering a regionally based interjurisdictional entity, the purpose of which would be to define and analyze the problem and develop specific objectives that would accomplish the regional goals. This entity would consist of representatives or stakeholders and the general public as well as technical experts. An example of such an entity is the Water Supply Citizens Advisory Committee (WSCAC) in the Commonwealth of Massachusetts (Holland 1996). Because representation in such groups is local, it may be necessary to devise different objectives for localities that play different roles in the regional processes of concern. In some cases, regional goals are developed by regional entities, such as in the case of the Great Lakes Water Quality Initiative.

Success for an EMSA depends on excellent communication among all involved. Thus, a general strategy for sustainable management of regional ecosystems (Figure 3) shows six spokes of a wheel, with each spoke indicating the type of information to be communicated among local, regional, national, and/or international stakeholders. The crosscutting issues and/or activities to be shared are goals, objectives, implementation, education, science, and monitoring.

Local implementation groups would be charged with devising plans to implement one or more regional objectives at the local level. Like the regional planning entities, these groups should consist of representatives or stakeholders and the general public as well as technical experts. However, the members of this group would be individuals who are closer to the problem than the members of the regional group. That is, they could include industrial waste managers, members of community environmental groups, local officials, and local resource harvesters rather than representatives of trade groups, national environmental organizations, and statewide agencies. They would use their knowledge of the local environment, waste-generating processes, natural resources, and political economy to generate a plan for meeting the objectives in a manner that is least costly and socially disruptive and most effective. The implementation plan would then be executed by the governmental bodies, industries, and nongovernmental organizations who participated in developing the plan.

In some cases, the system might work from the bottom up rather than from the top down. Local groups may perceive a problem that they believe is larger than the scope of their authority. Alternately, they may have been given objectives that they believe cannot be met at their level. In such cases, they may push the problem up to the national, state, or other governmental level through such traditional mechanisms as proposed legislation, petitions, and referenda. In other cases, it would be more appropriate to work through a regional planning entity, which could both devise region-appropriate objectives and coordinate activities of local groups.

Scientists play an educational and advisory role in this scheme. Each of the entities in this scheme must have sufficient understanding of the relationship between the needs of people in the region and the functioning of the regional ecosystem in order to devise realistic and efficacious goals, objectives, or implementation plans. In addition, each of the stakeholder groups must have sufficient understanding of ecosystem functioning over long terms and large scales in order to develop positions that reflect their long-term interests. This educational role could be served by universities, professional societies, consulting scientists, or staff scientists of the various governmental or stakeholder groups. Although it would be useful to develop materials such as articles or videos that explain general principles, this educational function must not be seen as equivalent to an adult education course. The educator-scientist must be willing to work with the groups to understand what they know and what they need to know.

The educational role of scientists gradually overlaps into the advisory role. Each entity in this scheme should have a scientific arm that could provide advice on technical issues, review plans and proposals for their feasibility and long-term consequences, conduct research needed to develop plans and objectives, and monitor the environment to determine the need for actions or the consequences of actions. This last scientific function is particularly important. One of the valuable lessons learned by environmental scientists is that actions have unintended consequences. Therefore, it is necessary to engage in an adaptive management process for environmental resources that monitors the environment and indicates when plans must be changed to reflect the reality of system response.

Clearly, many institutional impediments must be surmounted to ensure success of the EMSA approach, including the following:

- The short-term nature of planning cycles. Budgets in government and academia currently run in one-year cycles, with Congressional elections every two years. Only the most visionary politicians think in longer terms. Similarly, corporations tend to be driven by short-term profits, which ultimately may prove to be contrary to their long-term interests. To achieve SEM, much longer planning time frames are needed, e.g., five to ten years for near-term environmental planning plus a more visionary, long-range (50- to 100-year time frame) planning process.
- Inadequate public support for preventive rather than curative tasks. Prevention of environmental problems is usually cheaper than remediation, but results are difficult to quantify and needed actions are harder to justify. In contrast, curative or "command and control" action is expensive but results are observable. Because prevention successes are invisible, while remediation is noticeable, public support can be skewed toward exclusive interest in remediation. For this reason, active public education and outreach are of paramount concern in successful landscape and watershed management efforts.
- Stability in the face of sociopolitical change. When political offices change hands, agreements at the ecosystem or interjurisdictional level (such as ecosystem management strategies and plans) can be reversed or rendered inactive. The more formal or higher-level the agreement, the more successfully it will be transferred from one administration to the next. For example, formal state agreements, such as the Great Lakes Toxic Substances Control Agreement, have survived gubernatorial transitions because new governors inherited an obligation to their predecessors to fulfill their commitments. The longevity of an agreement's active life-span will depend upon the extent to which it is flexible and adaptive enough to incorporate changes in political priorities and inclinations. The Great Lakes Water Quality Agreement between Canada and the United States, which is high-level, is flexible at the local level, and has strong public support, remains active and visible.
- The makeup of the groups tasked with developing proposed agreements. Group membership that is principally technically oriented can produce the most scientifically sound approach but may not produce the most politically viable approach. Groups weighted too much toward policy-makers can ignore essential scientific realities, such that the proposal is successfully carried through but with limited benefit to the resource. It is important for groups designing SEM strategies to have within their ranks experts in both policy and science.

The success of the Great Lakes planning initiatives described in Chapter 3 (section entitled "Scientific and management issues for the Great Lakes") offers strong evidence that these obstacles can be overcome, provided that scientists become directly involved (and learn how to become productively involved) in the local and regional planning process.

Product Stewardship

A framework to implement sustainable product development (or Design for Sustainable Development [DFSD]) would carry out an organization's sustainable development policy and make use of available tools to develop the best option for action (Figure 4).

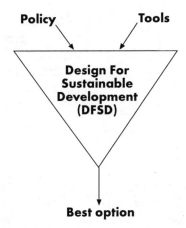

Figure 4 Framework for implementing sustainable product development

The DFSD framework consists of a variety of policies that describe the sustainable product development concept and a set of tools to implement them. Typically, the design of a new product (generically speaking, this design may include the development of a public-policy response) follows several stages, during which questions like the following must be answered:
- Objectives: Does the product fulfill the client's requirements?
- Internal capabilities: Does the product fit with the organization's core abilities?
- Competitive advantage: Can the product be brought to market at a competitive price?

This process is followed iteratively until the final design is achieved. In the case of a DFSD framework, the three key design criteria of environmental protection, economic health, and social responsibility (Figure 1, Chapter 1) are considered at every stage.

SETAC Press

Figure 5 illustrates typical energy and material flows for a product life cycle, from raw material acquisition to final disposal. A functioning DFSD framework should ensure that these issues are addressed during the design of products so that impacts of product manufacture, use, and disposal can be accounted for and minimized.

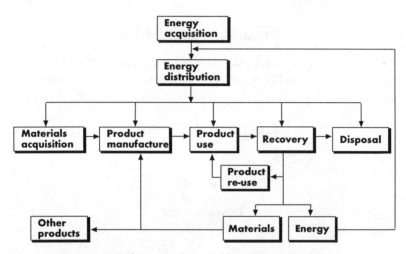

Figure 5 Typical energy and material flows for a product life cycle

One of the goals for a DFSD designer might be to optimize the use of materials and energy and to minimize their loss to the system. For example, a sustainable product strategy could be to ensure the collection of the material or product after it has exhausted its useful life span. Once the material or product has been collected, options for reuse, material recovery, and energy recovery exist. The specific option would be influenced by many factors, including the material itself, the ability of the raw material to be managed in a sustainable fashion, and its inherent heat value. For example, the importance of recovering used aluminum in beverage containers for use as feedstock to replace virgin mined bauxite was learned many years ago. Recycling produces energy savings for the aluminum industry of nearly 90% when compared to a linear one-time use of the aluminum. In contrast, some studies suggest that overall impacts from paper recycling actually increase after the recycling rate reaches a certain level. A more holistic approach to evaluating environmental burdens of product development is needed to ensure that unexpected environmental impacts are not caused by design changes intended to solve a single environmental problem.

Two existing industry-sponsored approaches to environmentally compatible product development provide examples of problems and promises associated with existing approaches: Design for the Environment (DFE) (Allenby and Fullerton 1991–92; Allenby 1994) and Responsible Care.

Design For the Environment

Design for the Environment is a systems-oriented approach for designing more ecologically and economically sustainable product systems. It couples the product development cycle used in business with the physical life cycle of a product. DFE integrates environmental requirements into the earliest stages of design so total impacts caused by product systems can be reduced. In the life-cycle design, environmental, performance, cost, cultural, and legal requirements are balanced. Concepts such as concurrent design, TQM, cross-disciplinary teams, and multi-attribute decision-making are essential elements of life-cycle design. Presently, DFE programs tend to allocate too many of their activities to schemes for recycling. Thus, in the future, more efforts must be devoted to expanding the work to encompass all components originally highlighted as parts of the concept.

Responsible Care

Responsible Care is a chemical-industry initiative attaining worldwide acceptance. It is designed not only to focus on safe chemical handling but also to improve performance in the fields of environment, health, safety, product safety, distribution, emergency response, and relations with the public. Responsible Care also enables companies to demonstrate that these improvements are actually taking place. The main thrust of the Responsible Care program is to achieve real improvement in the environmental, health, and safety performance of a company and to clarify who has responsibility for these areas in a given situation. However, Responsible Care is about actual performance, not merely public relations.

Challenges and priorities in the future include the following:
- development of environmental, health, and safety management practices and systems that ensure continuity despite changes in personnel;
- ensuring that all staff — engineers, technicians, supervisors, and managers — have required knowledge and skill; and
- encouragement of more open communications with companies both inside and outside of the chemical industry, to increase awareness, commitment, and the generation of ideas for improvement.

Potentially useful techniques

The following are examples of specific techniques that can be implemented within sustainable product development programs.

Environmental impact assessment

Environmental impact assessment (EIA) (Westman 1985) is a stepwise procedure to collect, organize, analyze, and evaluate necessary information about the character of a project, the environmental background conditions, the potential environmental effects in the future, the scope of consequences for people, and the need for remedial measures. EIA of project plans and proposals includes four major components: initial environmental examination, identification, prediction, and evaluation. Future development of EIA should focus on incorporating its sequential components into process and product planning to give necessary and early inputs to corporate planning and decision-making.

Environmental auditing

Environmental auditing (EA) is the process whereby selected levels of a company's organization are judged with regard to compliance with regulatory requirements and internal policies. It is a management tool comprising a systematic, documented, periodic, and objective evaluation of how well the environmental organization, management, and equipment are performing. The aim is to help safeguard the environment, health, and safety by facilitating management control of environmental practices and assessing compliance with company policies, including meeting regulatory requirements. The scope of an EA encompasses all types of activities that can adversely affect the work and natural environments.

A current limitation of EAs is that they tend to focus too much on existing regulatory compliance. To ensure that compliance with company policies (which often are product-related and future-oriented) is considered, product development must be examined in its totality, based on a life-cycle concept. Simple, qualitative, screening LCA procedures might be preferable to EA for this purpose.

Pollution prevention

Pollution prevention is any practice that reduces the amount of environmental and health impacts of a pollutant released into the environment prior to recycling, treatment, or disposal. Pollution prevention includes modifications of equipment and processes, reformulation or redesign of products and processes, substitution of raw materials, and improvements in housekeeping, maintenance, training, or inventory control.

Environmental labeling

Environmental labeling is a means of distributing positive consumer information. By using symbols, slogans, or other messages, companies can indicate to their customers that a product has certain advantageous properties from an environmental point of view. Environmental labeling of products can be an effective environmental protection tool if it guides consumers to purchase products that are friendly to the environment and if it stimulates industry toward environmentally sound product development.

Existing ecolabeling schemes are usually based on single-criterion approaches and, hence, do not take into account all environmental aspects of a product. One way to expand the scope of ecolabeling might be to introduce a multi-criterion approach and LCA, giving information to a variety of potential audiences about environmental impacts through the different phases of a product's life cycle, from raw material acquisition to waste disposal.

Life-cycle assessment

Life-cycle assessment evaluates the environmental burdens associated with a process, product, or activity by identifying and quantifying energy and materials used and wastes released to the environment, to assess the impact of those energy and materials uses and releases to the environment, and to evaluate and implement opportunities to effect environmental improvements. The assessment includes the entire life cycle of a product, process or activity, encompassing extracting and processing raw materials; manufacturing, transportation and distribution; use, re-use, maintenance, and recycling or disposal.

Presently, most LCA studies focus on inventory analysis and do not include impact assessments. Some LCA studies carried out in recent years have used various approaches to classification, characterization and valuation techniques. One such approach, the EPS, is described in detail in Chapter 3.

Full cost accounting

When businesspeople enjoy entrepreneurial freedom, the efficiency of finding new ways to meet consumer demands and generate quantitative growth has been clearly demonstrated. From an environmental standpoint, it is important to fully integrate the value of environmental assets into operations (much as the value of real estate is often included at present) in order to guarantee conservation of nature for future generations. Present market prices for energy and raw materials rarely reflect their direct cost, perhaps even less so if they are subsidized by governments. However, prices do not also reflect their true ecological costs (including environmental damage), nor do they consider the depletion of "natural capital" when a natural resource is used.

There is a need to internalize such external costs (i.e., costs for using nonrenewable resources, for environmental remediation, for cleanup of waste disposal sites, etc.) into companies' economic accounting. This principle, referred to as full cost accounting (FCA), is a managerial accounting method that assigns both direct and indirect costs to specific products. FCA is being trusted to make companies more aware of the entire environmental burden associated with their operations. FCA is still a new concept to many companies and, hence, has not been widely applied. Therefore, little experience has been gained about its advantages and disadvantages. Future adoption of FCA on a broad scale will indicate whether the concept needs to be further refined.

Risk assessment

Risk assessments evaluate the probability that human activities or natural disasters will have undesired effects on human health and the environment. General concepts and definitions relating to risk assessment have been published by the National Research Council (1983, 1993) and by USEPA (1992). All of these sources describe risk assessment as a four-step process. The following scheme, taken from the 1993 report of the National Research Council, is the most general in that it is intended to apply to both human health and ecological risks. The four steps defined in that report include these:

- hazard identification (determining the types and potential severities of adverse health or ecological effects associated with a contaminant or other hazardous agent),
- exposure assessment (estimating concentration and duration of exposure to a hazard),
- exposure–response assessment (estimating the response to environmental exposure), and
- risk characterization (combining exposure assessment and exposure–response assessment, evaluation of uncertainties, and presentation of results).

Presently, there are many technical limitations on the capability of risk assessors to quantify health and ecological risks (National Research Council 1993). Nonetheless, the concept of risk assessment as a bridge between scientific research and environmental management is generally applicable to sustainable product development, and many of the quantitative techniques can be integrated into LCA and other approaches for sustainable product development.

Making the transition: Ten steps to sustainability

The following are intended to be representative steps an organization might follow to fully integrate an SEM philosophy into its operations. This process (Figure 6) is intended to be applied to any organization (e.g., government, university, consulting firm, industry) that designs, manufactures, uses, and disposes of products, processes, and technologies.

Step 1: Appoint a sustainable development champion.

This champion should be someone who has the authority, visibility, commitment, and accountability to ensure that sustainability-oriented procedures are developed and implemented. Because clear support from senior positions in an organization is critical for implementing real change, responsibility and authority for such change might be assigned to a person at the level of corporate vice president or equivalent.

Step 2: Establish a sustainable development implementation team.

The sustainable development implementation team should have the commitment from senior management to develop and implement the plan. Given the all-encompassing elements of sustainable development, the team should include representatives from all relevant functions, e.g., management, research and development, marketing and sales, and human resources. This sustainable development implementation team would then be charged with the responsibility for developing the sustainable development implementation plan.

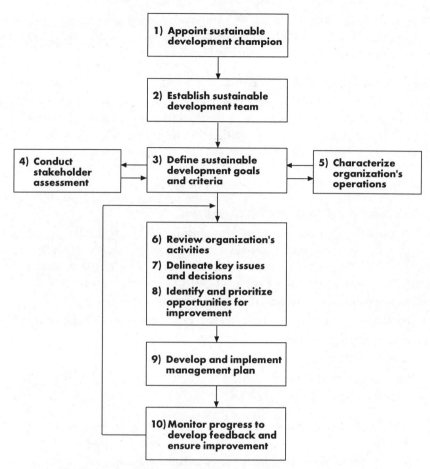

Figure 6 Integrating sustainable environmental management philosophy into organizational operations

Step 3: Define sustainable development goals and criteria in the context of the operation's business.

This might include the development of a mission statement and the establishment of values, principles, criteria, and goals related to profitability (long- and short-term), environmental protection, and social responsibility. Such management statements of sustainable objectives will likely address areas of economic, environmental, and social responsibilities of the organization. These goals should be based on information that is linked to organizational performance and that decision-makers can utilize to test the level of effectiveness of their sustainability efforts.

Specific operational goals within the broad mission could then be defined, e.g., implementation of FCA, consideration of public responsibility in product development decisions, or assurance of long-term resource availability.

Step 4: Conduct a stakeholder assessment.

The purpose of this step is to identify stakeholder needs and desires. Potential stakeholders include employees, board of directors, shareholders and investors, customers, regulatory agencies, suppliers, local communities, and the general public. Motives and goals of these diverse groups are highly variable and may, for many issues, conflict. The objective of stakeholder assessment is to identify these motives and goals and, to the extent feasible, incorporate them in the implementation of the plan.

Step 5: Characterize the organization's operations.

The purpose of this step is to characterize the organization's current and potential operations activities. These operations should not necessarily be traditional organizational functions; rather, they should be functions fundamental to the long-term success of the organization. The following are examples of such operations or functions:

- product and process research and development;
- purchasing, including materials and energy acquisition and management;
- manufacturing processes;
- communication and education;
- human resources; and
- finance.

Step 6: Review the organization's activities in the context of sustainable development policy.

This step involves critically examining the operation's existing vision, goals, decision-making criteria against the sustainable development policy and stakeholder needs. The ultimate output of this step would be a list of ideas and thoughts about how to begin to direct and manage the process of change, from the current organization strategies to the actions necessary to meet the sustainable development policy.

Step 7: Delineate key issues and decisions.

The sustainable development implementation team must identify and define the key issues and the major decisions associated with each of the organization's activities. For example, the planning operation might make decisions about facility site selections or new technology investments. In the research and development department, decisions might include how to design a new product. In production, decisions might be related to expanding or eliminating products or product lines.

Chapter 4: Steps toward sustainable environmental management

Step 8: Identify and prioritize opportunities for improvement.
There are many ways an organization can identify and prioritize opportunities for improvement. In general, the matrix of information required for a sustainable development analysis would focus on how key issue categories could be integrated as measurable criteria into the major elements of an organization's business cycle (e.g., planning, research, engineering, and manufacturing). A matrix provides a systematic way of gathering and integrating issues and values for each of the three sustainable development components: economic health, social responsibility, and environmental protection (Figure 1, Chapter 1). This information can then be used to identify common themes, gaps, missing systems, tools, data, and information needs that represent opportunities for improvement.

An example template for such a matrix is shown in Table 4. In each of the matrix cells, the planners would provide a measurable criterion for successfully addressing the key issue at hand during the appropriate portion of the business cycle in a manner consistent with the organization's sustainable development policies. Tools or information systems that might provide relevant data include LCA, FCA, DFE, and pollution prevention programs.

Table 4 Sustainable development strategic matrix

Activities	Key economic issues	Key environmental issues	Key social issues	Resolution
Strategic and product planning	1) 2)	1) 2)	1) 2)	1) 2)
Research and development (product and process)	1) 2)	1) 2)	1) 2)	1) 2)
Resource acquisition and management	1) 2)	1) 2)	1) 2)	1) 2)
Engineering	1) 2)	1) 2)	1) 2)	1) 2)
Manufacturing	1) 2)	1) 2)	1) 2)	1) 2)
Human resources	1) 2)	1) 2)	1) 2)	1) 2)
Communication and education (internal and external)	1) 2)	1) 2)	1) 2)	1) 2)

Step 9: Develop and implement a management plan.

Based upon identified opportunities, a new or revised management plan would be developed and initiated. The plan should include organizational policies, practices, and processes aimed at achieving sustainable development goals.

Step 10: Monitor progress to develop feedback and ensure continuous improvement.

This final step involves the establishment of monitoring, feedback, and follow-up actions to correct or improve the sustainable development implementation strategy or management system. This step is a specific application of TQM, a philosophy that is now widely accepted in the business community.

Implementation strategies and tools

In order to effectively achieve the goal of environmental sustainability, an organization will likely have to consider a host of strategies and develop or deploy an array of tools to achieve results. The transformation from current state to future vision requires the deployment of specific strategies and the utilization of specific tools and systems (Figure 7). Note that as an organization starts to implement parts of the vision, the number of tools and systems deployed increases. Two strategies are available to organizations wishing to transform their organizational culture: education and policy.

Educational strategies

In the area of education, an organization first should decide who is the target of the educational program. Possible focus areas include constituents/citizens, consumers/customers, employees, teachers, children, mass media, thought leaders, and generally any practitioner of SEM. In a private setting, these practitioners could include key individuals in senior management, design, engineering, finance, manufacturing, procurement, and human resources. Once the target audience is identified, the message will likely have to be tailored to that audience, building on previously successful initiatives. For example, strategists should consider positioning environmental sustainability as a natural extension of the current focus on recycling. Further suggestions include developing training programs for teachers, utilizing practical case examples, building education into school curricula at all levels, and organizing professional conferences. Whatever the path for the educational program, its developers should recognize the importance of this dimension in achieving the goal of sustainability and should provide the details in a manageable and digestible fashion.

Policy strategies

Like education, development and deployment of specific policies are effective ways for an organization to implement the vision of SEM. Within the broad area of concepts for sustainable development, a number of organizations have already suggested policies and principles to utilize during the transformation process. Actions that an organization might take to implement these principles include the following:

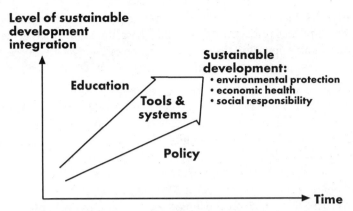

Figure 7 Implementing sustainable development strategies and tools

- Develop a statement of basic philosophy.
- Develop guidelines for action, possibly to include general management policies, corporate organization, a statement of concern for the environment, commitment to technology development, commitment to technology transfer, emergency measures, public relations and education, community relations, positions on overseas operations, contribution to public policies, and response to global issues.

Organizations should tailor the above suggestions to best fit their needs. It should be recognized that there likely will not be a one-size-fits-all set of policies. Once the policies are established, a host of implementation methods are available to any creative organization. These methods range from prescriptive to voluntary and from various incentives to potential disincentives. Again, the choice of implementation options is dependent on the organization's specific needs or goals.

Social Choice Mechanisms and Biodiversity

Maintenance of biodiversity is a prerequisite for the continued stability of human societies. Yet, because today's users of biological resources often have little or no direct incentive to preserve those resources for future generations, it has been difficult to provide for adequate protection and restoration of biodiversity.

Protection of biodiversity requires society to establish institutions and policies capable of directing decisions toward SEM. A variety of social choice mechanisms exist for this purpose, with varying capacities to effect such change. Among the most important of these are economies and governmental authority. Economic markets strongly influence private-sector behavior, for both individuals and corporations. Yet they often fail to promote efficient allocation of resources, including biodiversity-directed resources, for many reasons: imperfect knowledge, imperfect competition, biases created by governmental intervention, and an inability to consider economic impacts or provide public goods that fall outside of the normal price structure. Despite these limitations, market mechanisms may offer considerable opportunities in the protection of biodiversity. Success is conditional on finding ways to correct or compensate for the market failures noted.

Neither economic markets nor governmental decision-making provide well for the needs of future generations, whose interests are insufficiently represented by the present generation and command less attention and deference than intergenerational equity and long-term protection of biodiversity require. Governments rely on a range of policy tools, including public education, regulation of individual and corporate behavior, provision of taxes and subsidies (which may have either positive or negative environmental effects), and public ownership with appropriate public trust entitlements. Ideally, such intervention provides for the larger and longer-term public interest. In practice, however, governments often fail to perform well, and programs of biodiversity conservation are no exception. Public policies may be poorly designed; they may prove difficult to implement due to inadequate resources, conflicting goals and objectives, incompetent management, and insufficient public and stakeholder support, among other reasons; and conflicts may arise among agencies and between the federal government and other governmental entities. Much as market systems need to be improved, government must be reinvented. What is needed are better indicators of governmental performance, policy and program evaluations that provide useful and timely feedback to decision-makers, and more effective tools of public management.

Given both market and public-sector limitations, special consideration is needed to design mechanisms to establish and secure long-term commitments to protection of biodiversity. One mechanism is public trust entitlements, which are public-sector tools that can be, and have been, used to guide private-sector behavior when the marketplace is demonstrably not serving the public interest. Public trust entitlements are particularly appropriate for protection of land resources threatened by development that could put biological diversity at risk. They accomplish such protection by providing sufficient authority and a long-enough time frame to promote resource stewardship. Land ownership is assumed by government with compensation to property owners, and the property can be managed indefinitely to keep it in a natural state. The land is removed from the risk of development that would otherwise exist under any other public policy that is subject to change as economic conditions and human perceptions shift over time. The acceptability of public trust entitlements could be enhanced through use of an independent national commission to identify, evaluate, and select appropriate sites for inclusion in such biodiversity reserves. Public officials could then vote on a recommended package of acquisitions or land transfers that would free them from the political pressures attendant to choosing or rejecting any single site.

Other policy approaches may be attractive as well, including various public–private hybrids similar to the work of the Nature Conservancy. Such hybrids would benefit from the advantages markets offer for making individual choices and from the capacity of public sector institutions to formulate policies for protection of longer-term collective interests.

Creating long-term institutional commitments and public trust

Sustainable environmental management depends on institutions with capacities to protect biodiversity resources over time frames that are virtually unprecedented in human affairs. The successful operation of such institutions requires adaptation to changing conditions, rather than static management, and an ability to command long-term trust and support from individuals, corporations, and governments. Public trust and support are especially important to secure the commitment of SEM institutions to long-term goals and to guard against pressures to compromise those goals to provide short-term benefits inconsistent with protection of biodiversity. Insufficient trust undercuts the ability of the institutions to pursue their missions and, over time, could negate the best-designed biodiversity programs.

There exists some experience in the design of such institutions and of the policies and programs they implement. For example, public–private partnerships in creation of biodiversity reserves suggest the potential of combining the virtues of market systems with collective social choice. Other examples might be found in the work of soil conservation districts, water management districts, and similar enterprises. What is called for is review and evaluation of such experience to determine suitable institutional designs for the special case of biodiversity. Since existing economic, legal, and political paradigms often are not sufficient to meet these expectations, a wide net must be cast in search of appropriate institutional arrangements and processes that can assure public trust and support.

Historical analysis provides some examples, including continuity of the constitutional order in the United States over a 200-year period. Maintenance and adaptation of the structure of government and decision-making processes have been possible because of public support for basic principles of democratic governance and protection of individual liberties as primary social values. Changes occur over time without compromising these fundamental principles, which are an integral part of the social and political contract that created and sustains the U.S. political system.

Based on this and other experience in the United States, it seems likely that SEM will continue to rely more on governmental ownership or long-term contracts backed by the authority of government than on the private sector. The most striking example is the establishment of national parks and wilderness areas. Such areas are protected from development pressures under the legal authority of the federal government. The continuity of the policy is guaranteed as long as public support exists for setting aside such land. That support in turn depends on the public's conviction that such policies are in the national interest and benefit the people themselves. For national parks, public values are reinforced through opportunities to visit the parks and enjoy the amenities they offer. Support for wilderness areas is less direct and more symbolic but nonetheless present. It is sustained by providing at least some degree of public access to wilderness areas, albeit constrained by the necessity of maintaining the qualities of wilderness itself, which require severe constraints on the intrusion of human civilization.

Protecting biodiversity by creating nature reserves in which the public will be prohibited will be more difficult than establishing national parks and will require substantial effort to educate the public, stakeholder groups, and policy makers. Experience with the Endangered Species Act (1973) indicates the degree to which biodiversity goals may conflict with economic development interests, particularly when the scientific basis for preserving large land areas is not well established or understood by stakeholders. Resolving such conflicts and building consensus for protection of biodiversity will not be easy, but mechanisms for stakeholder involvement, education, and negotiation exist that offer promise over the long term.

A variety of other public policies and institutions may be envisioned that could offer long-term protection of biodiversity resources. Aside from creation of public trust entitlements, or public ownership of critical habitat, protection may be assured through creation of state or regional compacts with or without federal government involvement. Land owned by one department of government, e.g., the military, may be converted to a protected status and managed as a biological reserve where the qualities of the land warrant it. Another option would be to create long-term contracts among private parties, enforceable through established legal processes. Benefits may be established to provide incentives for signing such contracts, e.g., tax credits or tax reductions similar to those used for private land easements that serve a public purpose. On a local, state, or regional basis, government agencies may be empowered to constrain development of privately owned lands, much as has been done in California to protect coastal areas. This may be done in a way that is both fair to property owners and commanding of public support.

All such policies are invariably subject to political pressures from affected interests during implementation. And the policy goals themselves are open to reconsideration and revision over time. There can be no guarantee that future publics will not abandon the initial commitment to protect terrestrial, freshwater, or estuarine biodiversity resources. Although the same is true for establishing public trust entitlements, the latter are far less vulnerable to repeal. For all practical purposes, the question of their continuity is removed from the political process. In all these cases, however, deference to democratic processes and belief in the rights of each generation to make its own policy choices require that protection of biodiversity cannot and should not be removed entirely from the political agenda. Indeed, forcing each generation to revisit the question of the extent and form of biodiversity protection is a healthy component of democratic and sustainable environmental management.

Legal applications

Landscapes commonly consist of terrestrial, freshwater, and, sometimes, estuarine subsystems. Ownership law, property rights, usuary rights, and public-trust-doctrine precedents are dramatically different among these entities. For example, in the eastern half of the United States, organisms that occur in fresh water are publicly owned by virtue of sovereign-lands policy. Vertebrates, and increasingly many other organisms, also fall under public ownership because of precedential case law. Management of aquatic biodiversity (and thus aquatic site management) falls under state control because of public trust doctrines (e.g., water quality regulations, sedimentation control, dredging restrictions, etc.). But in the western United States, water law is based on private ownership, and generally the public trust doctrines do little, if anything, to protect aquatic biodiversity.

Terrestrial habitats can be totally owned by the private sector, and even the Endangered Species Act does not apply on these lands (unless malice is proven). Thus, the biological resources of the people of the United States fall prey to the "property rights" of the private sector, or in some cases to poor management by public-sector agencies. Although it is true that wildlife still belongs to the people of the state, access to it may commonly be denied by the owner. The management of biodiversity often falls through the cracks, inasmuch as

- it is "owned" by the people,
- its jurisdiction falls to the state, and
- any management that might occur represents charity from the owner of the land upon which it occurs.

Coastal marine habitat management responsibilities are more complex. Private habitat ownership protrudes into the sovereign waters of the public domain when it extends down to the low tide level. But in a time of rising sea levels, the "low tide zone" keeps changing, and it is unclear who owns what or has authority over what.

The landscape pattern of these habitat-specific, legally mandated jurisdictions and responsibilities must be superimposed on the dynamic spatial and temporal patterns of habitats and species. Currently there are several large federal programs undertaking the task of gathering information and elucidating these ecological patterns. The USGS Biological Resources Division, the USEPA's Environmental Monitoring and Assessment Program (EMAP), the U.S. Fish and Wildlife Service's GAP, and the Nature Conservancy's state heritage inventory programs represent the most notable examples.

Once this GIS-based profile of national biodiversity resources becomes available, those public domain lands that are biodiversity and/or ecodiversity process "hot spots" could be designated as management foci. These biodiversity foci might achieve a social standing at least as high as national parks, monuments, museums, or archives.

Severed property rights, i.e., the condition of split ownership, require special consideration. For example, situations in which control over the aquatic component lies in the public sector, but the riparian zone lies with the private sector, require a combination of entitlements and performance-based programs (Best Management Practices, voluntary codes of ethics, creative zoning and ordinances, etc.).

A reconfiguration of public domain lands might well be the single greatest step that could be taken to reconstitute presently fragmented and dysfunctional ecological systems. Political boundaries are rarely congruent with ecological boundaries, and large gains in biodiversity conservation could be made by land trading between and within the private and public sectors. Because this recommendation seems so critical, a specific example of what it implies is cited: As of the mid 1980s, the U.S. government had declared a Native American school property (located in a prime business development setting in the Phoenix, Arizona area) to be surplus government property. Simultaneously, additional, privately owned land was needed to expand the U.S. National Park Service holdings in south Florida. Creative entrepreneurial activity allowed the swap of a large parcel of badly needed ranch land for a much more economically valuable surplus school property. This exemplifies the myriad win–win opportunities that occur all over the country as more is learned about the location of critical hot spots of biodiversity and equally critical, important ecological process areas.

At present, about 40% of the U.S. land base lies in the public domain as national and state forests, national and state parks, military bases, or privately owned biodiversity preserves. Yet, there still exists a consistent loss of native diversity. Simply buying back more private land is not the answer. Reconfiguring the ownership is more likely to achieve results. Landscape ecology-based, performance-based, incentive-driven new programming that utilizes innovative approaches such as land swaps, land trusts, mitigation land banks, and debt-for-nature allowances is more likely to achieve results.

Private-sector management is usually driven by perceived market values, and yet our ability to calculate dollar values for biological diversity is hopelessly under-researched and naive. Intergenerational equity (sustainability) is a public-sector commitment, and this obligation is not to be entrusted solely to private-sector approaches. Further integration of the complementary strengths of the private and public sectors must be pursued.

Ensuring Sustainability of Urban Environments

Ryding (1992) has effectively presented the issues of environmental management as being aggregated into broad sectors of concern: agriculture, forestry, industry, energy, transportation, and the urban environment. This last sector encompasses health care, nutrition (food processing and distribution), industry (employment and economic vitality), housing, communications, water supply, energy, recreation and open space, and, especially, waste management. Each sector can be represented as a series of flows of mass, energy, or human activity; most interact in the ecological sense and certainly in the system sense of a landscape or higher level of biological organization (Figure 8). The consequent complexity has presented serious problems in understanding and analysis, yet often can be measured and evaluated by fairly simple methods. The result is the perceived view of a city region as an organismic analogy involving uptake, metabolism, storage, and excretion of materials.

Chapter 4: Steps toward sustainable environmental management 63

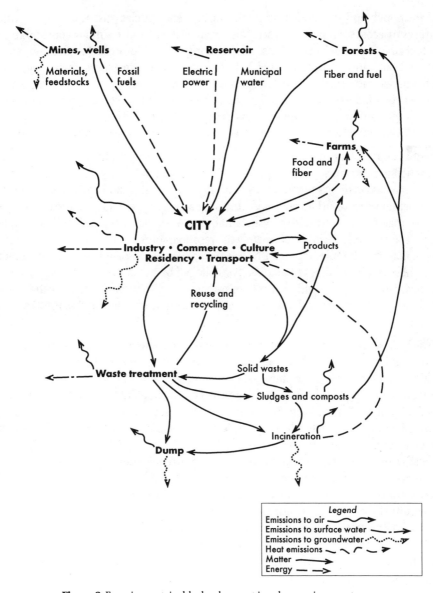

Figure 8 Ensuring sustainable development in urban environments

For any population and biotic community, density and age class distributions are critical features of ecosystem performance. Sustainable cities require similar considerations. "Balanced indigenous populations" of the young and old, of extended families, and of different skills and interests are vital. Within any set of circumstances, there may be limits to the size and density of the total population sustainable by a city and its regional resource network. Determination of how these densities will affect the function of the city are as important to that system's success as similar data would be for a fishery or wildlife area. Development within each of the sectors of concern requires such consideration.

Science is not just a compilation of facts and process descriptions; it is the disciplined approach to problem solving, hypothesis testing, and analysis by deduction and inference from known or accepted theories. The application of natural and social sciences to the several sectors of urban environmental management presents a superb opportunity to take major steps toward urban sustainability. At the same time, the process underscores the principle that it is not merely the resources or the environment that are actually managed, but more importantly, it is the actions of the people who use or interact with these resources that are managed or changed by use of science. In each sector of concern, groups of scientific disciplines are necessary to understand allocation of resources, investment of time and money, and application of technical skills and methods to problem resolution.

Sustainable approaches to urban environmental management may best be characterized as stakeholder-originated, bottom-up, cooperative, multi-media, integrated processes that account for interrelationships, establish linkages, and take actions that lead to sustainability. One potentially successful approach is exemplified by the RAPs being developed for polluted rivers, harbors, and embayments on the Great Lakes (Hartig and Law 1994). The steps in the development of RAPs (Figure 9) involve close cooperation between interdisciplinary teams of technical experts working with stakeholders and responsible officials. Arbitrary application of standard regulatory criteria or technologies is replaced with integrated permitting of air, water, and land use based on site-specific analysis of the problem. The cornerstone of the RAP process is shared decision-making in a multi-stakeholder institutional framework tailored to the local area. Multi-stakeholder RAP groups are reaching agreement at key points in the decision-making process. Progress is sustained by an iterative process of establishing short-term priorities and celebrating milestones (Hartig and Zarull 1992). The result is action planning within a strategic framework embodying community accountability.

Possible regulatory and technical approaches

Conventional environmental management has basically failed to make the city environs livable, much less enjoyable and uplifting in the manner often envisioned by those attracted to them. Many problems, however, are not the fault of environmental management per se, but rather reflect the disconnected management of the entire complex of urban or suburban features. Addressing all of these areas across the board is a very daunting task; however, the management "sectors" identified by Ryding (1992) share

Chapter 4: Steps toward sustainable environmental management 65

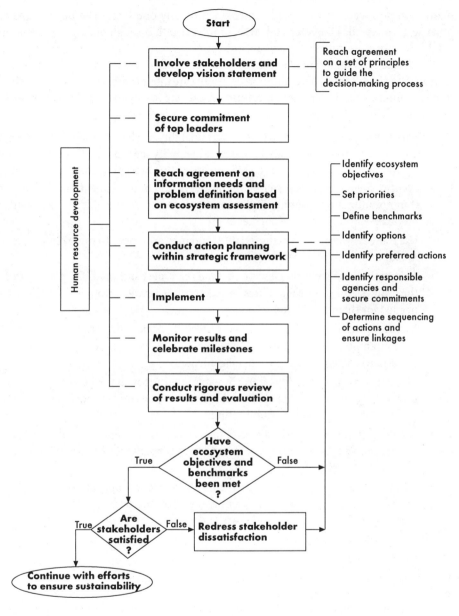

Figure 9 Development of a remedial action plan

many common features. Accomplishments within any one sector can be reflected in other sectors as well. Thus, steps taken to improve health through pollution prevention would also improve the aesthetic quality of the environment and lower costs of health care and structural maintenance (e.g., cleaning or repair of buildings and monuments damaged by air pollution). Organization of health care into neighborhood teams or community centers would increase opportunity for improved parenting skills and better health care delivery to babies.

Past experience shows that progress toward sustainability can be effectuated when incentive-based mechanisms of negotiated regulatory flexibility reward private initiatives that develop public benefits. Steps required to implement this process include the following:

- Identify the means to transfer planning and action or implementation authority (but not ultimate accountability) from appointed or elected officials to community-based, public–private partnerships.
- Use these new groups to define goals, standards, and measurement endpoints to manage change and to resolve disputes.
- Ascertain which technologies and methods can meet sustainability goals within the group's economic capabilities, and determine private or other incentives needed for their adaptation.
- Monitor progress toward sustainability, feeding back information to provide accountability, to create trust and effective communication between stakeholders, and to encourage adaptive management.
- Phase in new sustainability efforts by experimenting with low-level incentives or impacts to assess potential applicability for reducing conventional management and as training for private–public groups.
- Create single-source service agencies to provide small businesses with necessary permits and educate them about ways to minimize adverse environmental impacts.
- As the community gains experience, education, and training, expand the complexity of the group's tasks, identify scientific and technical needs for research and development, and share and celebrate successes.

Throughout this process, the roles of science and scientists (encompassing engineers, technologists, and applied social scientists) in leadership and support are apparent. In particular, viewpoints from environmental science, ecology, and systems management offer an objective currency for the standards and metrics of organizational performance. Ecological and health risk analyses in appropriate forms already can help prioritize and categorize environmental threats. Assessment of impacts and of the means applied to alter or prevent them can similarly be evaluated by a variety of methods embodying statements of uncertainty and statistical robustness. Even where such tools for a specific task may not exist right now, the means of obtaining them lie within the framework of SEM. Research, especially in regard to information management as well as environmental and ecological sciences, is likely to be a nonexclusive source of these tools.

New Approaches to Environmental Restoration

Restoration of ecosystems damaged by mineral extraction or industrial activity is an important component of any long-term strategy for SEM. Environmental restoration can be broadly categorized as remediation, reclamation, and planned restoration of ecological resources through site regeneration. Remediation includes those activities undertaken primarily to clean up contaminated or otherwise degraded environments and restore them to acceptable conditions. What is acceptable is determined jointly by regulatory agencies, the potentially affected public, the responsible party, and governments, both federal and state. Risk reduction and regulatory benchmarks routinely determine remediation goals.

Reclamation returns degraded ecological systems to their pre-disturbance conditions following natural- or human-induced degradation. Emphasis is placed on reestablishing ecological structure and function commensurate with the surrounding landscape. The goals of reclamation are determined by evaluating the ecological data about a site and taking action that is designed to restore habitat and inhabitants to the way they were prior to disturbance.

Restoration site planning can be interpreted broadly to include predetermined reclamation activities as part of planned alteration, use, or degradation of ecological resources. That is, specific plans for restoring the resources may be developed prior to resource degradation. Restoration might occur following the planned resource use, as in the case of remediation or reclamation. Alternatively, restoration activities might proceed in concert with scheduled resource use in an iterative manner, e.g., in the long-term use of renewable resources.

All three categories of restoration activity can be guided by the aims and goals of SEM. Under current practices, remediation and reclamation do not necessarily address issues of sustainability. With sustainability as a driver, the goals of environmental management in the private sector would extend beyond mere compliance. The management objectives would be consistent with long-term viability of ecological resources of concern. The costs would be, within reason, whatever it took to realize the objectives. The costs of restoration under a sustainable management program would likely be no greater than (or might even be less than) those of conventional restoration, when the benefits of less costly, long-term maintenance are considered.

The principal differences between conventional and sustainable environmental restoration strategies are summarized in Table 5. SEM is holistic in outlook, while conventional environmental management takes a narrow focus. Sustainability-based environmental management actively addresses the social, political, and economic aspects of management and restoration, in addition to embracing a comprehensive ecological approach. Conventional management narrowly bounds its problems and addresses them in ecological and managerial isolation.

Table 5
Comparison of conventional management and sustainable environmental management methodologies

Conventional management	Sustainable environmental management
Narrow	Holistic
Single resource	Overall ecosystem, structure, function
Ownership	Stewardship
Command-and-control	Self interest-motivated
Reactive	Preventative
National goals	Local or regional goals
Linear solutions	Integrated solutions
Risk averse	Risk taking
Rigid	Adaptive

Conventional management is based principally on rights and responsibilities of ownership. Commonly, the owned resources do not constitute an ecologically sustainable unit. Managing parcels of land smaller than a self-sustainable unit requires the owner or manager to consider the ecological services required by the remainder of the system. Yet, the owner of one parcel cannot control the uses of others, nor does he or she have the financial resources to support them. Pesticides may be used to replace normal predator–prey interactions for cost reasons. Fertilizers may be needed to augment disrupted nutrient systems but may not be supportive of the entire system. If other owners of adjacent lands within the watershed have different management objectives, the efforts of the owner of the small parcel to realize her or his objectives may be ineffective or harmful. In contrast, management for sustainability emphasizes stewardship in place of ownership. Management of individually owned land parcels would be coordinated and relate directly to natural ecological units.

Sustainable environmental management relies on flexibility and adaptability in the management process. Decisions are viewed as experiments: monitoring the results of decisions produces information that can be used to refine, modify, or replace management practices in specific applications. Consequently, practitioners of SEM learn from previous successes and failures.

Under a rigid command-and-control philosophy, conventional environmental management tends to be reactive. In contrast, SEM is actively designed to anticipate and avoid problems. Conventional and highly centralized management tends to be risk aversive in its decision-making and regulatory activities. Currently, risk-taking characterizes SEM philosophy and approaches because mistakes are viewed as opportunities to refine the management process.

Conventional environmental management is highly centralized, particularly in the government sector. This centralization results in decision-making that occurs distant from the location of the problem. Such centralization tends to contribute to fragmented approaches to management, as well as to the inflexibility and command-and-control philosophy previously outlined.

Resulting decision and regulatory criteria are typically national in extent. SEM argues for moving the regulatory and decision criteria as closely as possible to the realized or potential problem. Local or regional solutions to problems are emphasized; decision criteria and management actions are purposefully specified in relation to the relevant ecological, sociological, economical, and political scales commensurate with the problem.

Conventional management employs a linearized system of resource utilization. Raw resources are collected from their source, manufactured or processed, distributed or sold, used, and discarded as waste. The U.S. Resource Conservation and Recovery Act (RCRA) of 1976 attempts to minimize the environmental impacts of a linear approach to resource utilization, but does little to change the economic forces underlying resource utilization and waste disposal. In contrast, an SEM approach would attempt to introduce the feedback mechanisms characteristic of naturally sustainable ecosystems to break as many of these linearized flows as possible. High ecological costs of resource extraction or waste disposal would "feed back" on industries responsible for the impacts, leading to the adoption of less destructive practices.

Another contrast between conventional and sustainable environmental management pertains directly to the ecological goals of restoration. Conventional restoration may attempt to reestablish a "snapshot" of the natural past, so that emphasis is placed on restoring structural similarity to pre-disturbance conditions. The composition and relative abundance of plant and animal species representative of similar, undisturbed systems are used as guides for what the restored system should look like. Restoration from a sustainable viewpoint would address function in addition to structure. Consideration would be given to the successional dynamics and restoring the potential for change. Additional thought would be directed toward different possible and likely future land use in designing the structure and function of the restored system. Local or regional economies would also be examined in relation to restoration activities. Flexibility for future alteration of the restored system without resulting devastation would be a sustainable management goal. Sustainable restoration would integrate the stakeholders' view as to the importance of naturalness, economic planning, and perceived human needs as part of the overall restoration plan.

Finally, under conventional management practices, motivations are different for the public and private sectors. With sustainability as the focus, these differences would decrease. The objectives, costs, and time scales of public-sector SEM would become the same as those in the private sector. The objective would not be restoration to a given national standard, but restoration to a point that the environment can renew itself by natural processes. Industry, government, and the public at large all would be responsible for ensuring this happens and, as necessary, for sharing the costs. The needed level of restoration would be determined locally by a process that involves all of the affected parties. The industry mind-set would view both extraction and restoration as parallel goals. The affected community and government would be viewed as partners in the process. The industry, the government, and the community would benefit when revenue is generated; likewise, they would benefit when the restoration is completed. Thus, instead of being a cost to industry, restoration would be a benefit, the planning process to reap this benefit would begin at the onset, and the social and financial responsibility would be shared by all of the participants as the project reaches fruition.

General principles for implementing sustainable environmental management

Conventional management strategies in the area of restoration of environmentally impacted land and water resources have not been adopted at random. Rather, they exist as the result of rational profit motives, regulatory inertia at both the federal and state levels, and insufficient information in general regarding the need for SEM in the restoration context. The profit motive of businesses is both expected and necessary. A business will restore lands as mandated by government to avoid a fine and at a cost consistent with a reasonable rate of return. No rational businessperson will adopt environmentally sustainable management practices unless doing so is consistent with this underlying premise of operation. Thus, to encourage change, any solution must be at least partly based on economic incentives that make both the environment and the business person better off. Examples of such incentives would be tax credits, reductions in liability for adoption of creative and flexible techniques of reclamation and restoration that promote sustainability, and promotion of public–private partnerships to help share costs.

A command-and-control strategy for implementing sustainable environmental management

One strategy for implementing environmental restoration in the context of SEM is to require sustainable restoration planning during the permitting activity for a new project. Reclamation plans are currently a required component of some permitting activities, e.g., when a new mine is permitted or a forest harvest plan is approved. These plans are typically focused on the stabilization of an area and the rapid achievement of specific environmental goals, such as the establishment of ground cover to prevent erosion. While reclamation goals are important and may be part of a sustainable restoration program, they are not sufficient for creating a self-sustaining environment. A sustainable restoration program would include all the components required for creating an environment capable of sustaining itself, be it a natural one modeled after what was in place prior to disturbance or a planned environment to meet specified human and ecological needs.

A possible sequence of events for this implementation strategy is illustrated in Figure 10. Once a user makes a decision that will result in disturbance or destruction of habitat, the appropriate regulatory agencies would be notified. The user would request a meeting

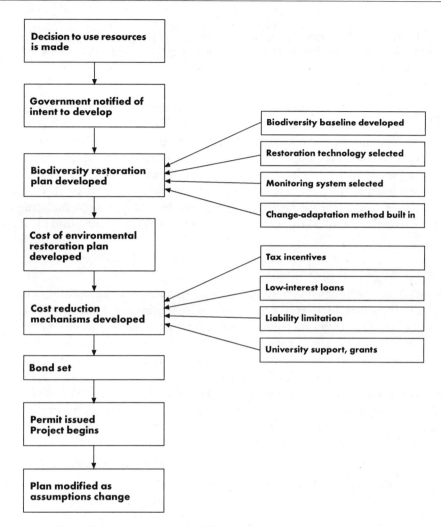

Figure 10 Incorporating sustainability principles into site restoration planning

with the agency to discuss the basic components of the project and to provide an assessment of the restoration requirements. The restoration plan would be based on an inventory of flora, fauna, and ecosystem processes. The restoration plan would include appropriate technologies and proposed monitoring systems and would be flexible enough to ensure that new conditions or considerations can be addressed as the project progresses.

The restoration plan also would include a schedule for implementation and a description of estimated costs. The cost estimate would provide the basis for the regulatory agency and the user to negotiate the size of bond that would ensure restoration after the resource extraction or utilization was complete. Upon completion of the permitting and placement of the bond, the regulating agency would issue the permit and the project would begin. As the project progressed, changes in the project that affect restoration activities or cause an unexpected effect on surrounding habitats would be addressed by modifying the restoration plan.

In order for this strategy to ensure successful, sustainable restoration plans, the resource developer and the regulatory agency must see themselves as partners in achieving responsible development of the resource. This partnership requires flexibility on the part of both partners and may include mechanisms for reducing the cost of the restoration program. Cost reduction mechanisms could include tax incentives, low interest loans, liability limitations, and opportunities to obtain university expertise in the form of applied research. Cost might also be reduced by adopting less stringent restoration criteria, particularly if the site will likely be exploited in the near term or if future land-use planning might justify less pristine conditions.

An adaptive planning strategy

Regulations for environmental restoration have been mandated by legislation in many countries. These regulations ensure that aspects of the environment that have been perturbed by human activities are replaced, but such regulations are generally too rigid to allow implementing flexible alternatives. Allowing more flexibility could lead to enhanced sustainability by developing goals that address issues beyond site boundaries. Regulations should therefore allow responsible parties to be creative in design and adaptive in execution of restoration activities. Such regulations would establish the framework and essential ecological elements to be considered, without being prescriptive as to technology or timing. This latter point is especially important because natural or ecological technical approaches may require more time for resolution.

Key to the execution of such regulations would be the building of a relationship of trust and partnership among the stakeholders, the restoring party, the surrounding community, and the regulators who represent them. Such a partnership would demand openness on the part of all participants and recognition of co-ownership of the restoration project. The partners would work together to define the level of ecological restoration that would satisfy principally local community needs while not conflicting with larger regulatory objectives.

Primacy of the "local solution" is a key element to assure that national objectives, which may be wholly inapplicable to site-specific conditions and inconsistent with sustainability interests, are not rigidly executed. Thus, the community that must live with the results of the restoration will have a major influence on the outcome of the process. A community and the industries that support its economic basis should be free to seek creative solutions within the context of the regional environment and within constraints imposed by pertinent, resource-specific regulations (e.g., Endangered Species Act). Economic incentives or creative social investments, as described in preceding sections, can provide solutions that benefit all the stakeholders.

Control at the local level would ensure that the people who must or will live with the environmental and economic results of the planned activity would have substantial input into activities that affect their environmental future. Overlaying a TQM philosophy would incorporate the intellectual construct of continuous improvement, which in turn will provide for adaptive management as the project develops. An adaptive management strategy (Figure 11) that moves from objective goal-setting through strategy selection and implementation could monitor, evaluate, and improve its approach to achieve the goals of the restoration project in the most efficient way.

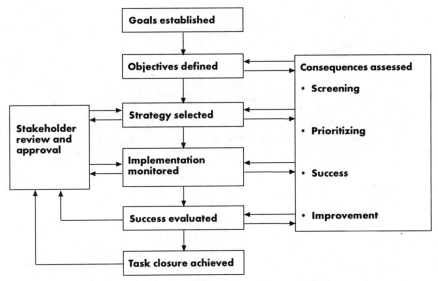

Figure 11 Adaptive management strategy

The regulatory documentation of such an adaptive approach (Figure 12) would include a clear definition of the planning process, stakeholders to be considered for inclusion (Table 6), necessary documentation, and the general or ecosystem-specific environmental quality objectives for planning.

If the intended operations are not the result of an accident or past practices, then the planning process would include five distinct phases:
1) Pre-disturbance planning: During this phase, the company or other entity develops a preliminary plan and then presents it to the local community. This plan should contain the basic elements for the assessment of environmental impacts, including project description, potential environmental impacts, mitigation procedures, and a restoration plan that addresses environmental sustainability. The company would then request a local stakeholders' review and participation in the completion of a plan for project implementation.

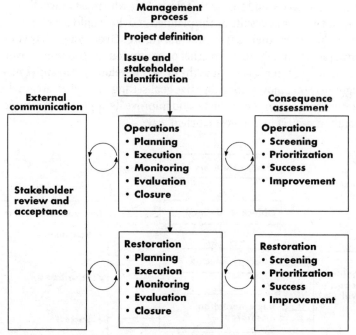

Figure 12 Adaptive management process for flexible restoration planning

Table 6 Potential stakeholders in adaptive restoration management

Primary	Secondary
Site management	Corporation management
Contiguous landowners	Corporation stockholders
Affected industries	Politicians (national)
Local government	Conservation organizations (national)
Local citizens	Trade organizations
State or federal regulators	

2) Operations: A positive result from the pre-disturbance planning phase would yield approval for operations and finalization of an initial plan of operations. Such a plan should include periodic monitoring of the consequences of operations, the data from which can be used to evaluate the level of operational success and to suggest areas for improvement. During this phase, periodic reports on the status of the operations could be filed for stakeholder review. Adaptive management would allow for modifications in operational strategy that could be proposed in the periodic report. Input from stakeholders would occur through periodic meetings with company representatives.

3) Operations closeout: Near the termination of operations, the company would notify the local community that it intends to close out operations in an area and initiate restoration. At this time, the preliminary restoration plan developed in the pre-disturbance planning phase would be reassessed and revised, as appropriate, based on the knowledge obtained by monitoring operations and by eliciting feedback from stakeholders. A revised restoration plan would be presented to the local community, and local stakeholders would conduct a final review. A final plan for restoration would then be developed and included in the final operations report.

4) Restoration: The restoration plan would be adaptively managed. Periodic monitoring of restoration activities would occur and be used to modify the restoration plan. Periodic status reports would be developed and made available for stakeholders' review. Stakeholders could request a meeting to discuss identified issues. As restoration comes to a close, sign off would be required. A draft post-restoration plan would be developed and presented to the stakeholders.

5) Post-restoration: The duration of this phase will be dependent on the technology selected for restoration. If a natural recovery option is selected, the physical operations of restoration may be completed quickly, but the actual recovery of the environment will be some distance in the future. Therefore, the post-recovery plan should include follow-up monitoring and periodic reports to and meetings with the stakeholders. Adaptive management approaches would be used to make any needed adjustments in restoration, based on the monitoring results.

The openness and transparency of the described process should lead to trust and partnership among the stakeholders. All participants should leave the process with the feeling that their issues were sincerely considered in the selection of the final technology. In principle, both the local community and its industrial partner should have a measure of control over their own futures. Although the process might be more open than the way in which some industries traditionally operate, the flexibility in selecting the most cost-effective technologies and the ability to negotiate other financial incentives should leave them with an improved financial bottom line. Stakeholder involvement would produce the benefits of local insights into concerns, values, and options for environmental restoration at the appropriate scale.

Economic incentives

Industries often engage in the process of planning a project that will require significant environmental restoration. Alternatively, they may be the owners of land in need of substantial environmental restoration because of their past activities. In the case of the project in the planning stage, they may decide to abandon the project altogether because the costs of restoration are too high. In the case of the need to restore previously disturbed lands, they will be forced, for economic reasons, to fight paying their restoration costs because the cost of the fight is less than the cost of restoration.

There are many reasons for this behavior. For example, the company that has extracted the resource may not be as efficient at cleanup as it is at extraction, may not have the cleanup resources because of economies of scale, or may not see the advantage of cleanup, other than to avoid a fine. There may be, however, companies that specialize in restoration that would welcome the task because they can do it at lower cost and can see a potential use for the restored land. A system could be developed through which the company responsible for the cleanup could convey the disturbed land (and the responsibility for restoration) to another at an agreed upon price.

If the new owner of the land would agree to use sustainable environmental management principles in restoring the land, then that owner could be given financial incentives to make the purchase and assume the liability. Examples of possible incentives include tax incentives to offset against future income, ad valorem tax relief (for a fixed period), or perhaps investment tax credits that could be credited against income from other businesses conducted by the same company.

Another economic strategy would be to provide noncash economic incentives. One such incentive might be permit credits in exchange for voluntary remedial and/or restoration activities. For example, the restoration of an abandoned industrial site for which no previous operator or waste generator can be found can present regulatory agencies with expensive responsibilities for investigation and cleanup. In some cases, the expertise and logistical ability to clean up and restore such a site may reside with a locally operating industry. While the industry's management may want to be good corporate citizens by assisting in rectifying the situation, often they will need some type of incentive to make such activity pass the scrutiny of the company directors and stockholders. Such incentives might take the form of regulatory "credits." For example, the company may agree to clean up a site using SEM principles. In exchange, the regulatory agency would reduce the time required to permit a new process modification. Community recognition of a company's restoration work is also a powerful incentive. Under this barter system, the operator nearest the job, who could do it in the most efficient manner, would perform the task in exchange for effective noncash incentives.

Restoration might be economically beneficial to the principal responsible party under limited circumstances. If the anticipated net worth of the restored resources exceeds the cost of restoration, a profit-driven plan for restoration may be viable. Where profit is unlikely, other kinds of economic incentives might be developed and implemented to stimulate sustainable restoration of degraded resources. As previously outlined, these incentives might take the form of tax credits, low interest loans, reductions in liability, and cost sharing in the public or private sector. In some cases, novel solutions might recommend themselves. For example, obligations for restoration might be sold to investors who would assume the responsibility and liability for restoration under conditions that they would also buy the rights to future use of the restored resources or receive a percentage of the gross proceeds generated by future development. The objective would be to shift the economic burden to the larger scale of the sustainable unit where the economic burden at the site level becomes an asset at the larger scale.

Establishing restoration priorities

The imbalance between the magnitude of current environmental degradation and the resources available for restoration require that priorities for restoration be established. These priorities should reflect the objectives and principles of sustainability. Priority setting must recognize the natural dynamics of ecological systems. Restoration through sustainability attempts to put the degraded system back into its natural dynamics or successional context. Restoring natural habitats must be given higher priority than simply focusing on reducing chemical contaminant concentrations. Reestablishing the capacity of degraded lands to capture solar energy inputs, recycle essential nutrients, and decompose organic matter should take precedence over maintaining particular species composition at the expense of unnatural energy or material subsidies, e.g., designing with nature instead of gardening horticultural exotics. Under this ecologically functional approach to setting priorities, emphasis is placed on restoring populations at lower trophic levels before attempting to reintroduce tertiary consumers. This prioritization would help to avoid future problems similar to the insufficient forage fish base for Great Lakes salmonids and the associated economic costs, both in lost revenues and increased monetary needs, to build forage fish hatcheries. Ultimately, the system would feed itself rather than be fed in perpetuity by humankind.

Transfer of information and technology

Sustainable environmental management should be the most cost-effective environmental management methodology for industry. However, environmental managers may not have the information necessary to appreciate the range of economic and public-perception advantages provided by operating a facility that strives to meet the objectives of SEM. The needed information may exist; the key is transferring it to the appropriate people. Taking advantage of the opportunities for environmental management is best accomplished by strong top management involvement as well as involvement of day-to-day operators in identifying opportunities such as those for waste reduction and process improvement. Providing systematic training within and across organizations and cross-disciplinary training among environmental scientists, engineers, and managers are important components in encouraging innovative improvements to facility performance.

Another method to improve the transfer of technical information regarding SEM would be to establish a database of pertinent information that is readily accessible to public- and private-sector users. Such a database might include ecological information on regional or local habitats and areas of particular interest or concern, applied chemistry and toxicology information, land and water resource information, process technology options, environmental technology options, and technical experts. The freeing of intellectual property would also increase the ability of facilities in developing countries to implement methods and technologies that would help them reach SEM more quickly.

A mechanism is needed for rapid and effective application of modern ecological theory to field environmental restoration situations. A means to promote the communication and transfer of theory and practice among theoretical ecologists, landscape ecologists, landscape designers, horticultural and agricultural scientists, systems scientists, and ecologists in general is urgently needed to improve the efficient transfer of state-of-the-art science to implementation in the field. Workshops, special symposia, new publications, special private-sector–government cooperatives or task forces, and nongovernmental organizations might all serve beneficially in this capacity.

An organization is needed to collect, implement, evaluate, and disseminate ecological models, parameter values, and validation data that might be used in sustainable restoration design. A systems approach to environmental restoration suggests that ecological models will play an increasingly important role in the design of specific remediation, reclamation, or planned activities (e.g., Starfield and Bleloch 1991). Ecological models have proliferated throughout the primary and secondary literature. A central organization or clearinghouse charged with collecting these models, along with parameter values necessary to execute them, could greatly facilitate restoration consistent with the aims of SEM. The organization might also be charged with evaluating and standardizing the models, as well as with collating an electronic database of information relevant to site-specific implementation of the models. The organization could be funded in part by those who use the models and in part by public and private grants.

Local, regional, or national GIS should be developed or coordinated to provide spatial values for data and parameters relevant to ecological restoration. These information systems should be electronically accessible. Physical, chemical, and biological information relevant to sustainable restoration has been and continues to be collected by corporations, individual researchers, local and state government, and various agencies throughout the federal government (e.g., USEPA, Department of the Interior, National Oceanic and Atmospheric Administration [NOAA], U.S. Geological Survey [USGS], U.S. Fish and Wildlife Service, U.S. Department of Energy [DOE]). Certain activities are underway to coordinate, publicize, and disseminate these data (USEPA 1992). Additional efforts should be directed at collecting and collating data and databases in easily accessible form for use in site-specific restoration. The USEPA's EMAP might serve as a model for constructing a parallel organization with the express mission of constructing the necessary GIS capabilities to support sustainable approaches to environmental restoration.

Overcoming regulatory inertia

In many countries, regulatory agencies rely on command-and-control mechanisms and the enactment of rigid restoration regulations with narrowly focused goals. This inertia must be changed to involve all stakeholders in the decision-making process and permit flexibility in determining the goals, methods, and timeline for restoration if SEM is to be incorporated into the process.

Relaxation of rigid regulations that focus on restoring site characteristics would permit a shift to broader environmental considerations and allow for the enhancement of desirable ecosystem characteristics on a local or regional scale. Additional approaches, such as economic incentives, could be combined with more flexible regulations to provide a better basis for sustainable management of restored ecosystems.

Chapter 5

Conclusions and Recommendations

This report does not suggest any radical new ideas nor does it suggest any new or great burden on government or industry to do something that is morally or politically "correct." Rather, it is premised on basic ecology. In order for any ecological system to be maintained over time, there must be both balance and parallel growth between and among all of the organisms surviving within that system. This balance, producing healthy regeneration, relies on this principle: Among the actors within a system and those acted upon, there must be a mutualistic relationship whereby all are better off, or at least no worse off, because of the actions each takes. That balance is not achieved quickly; it involves a process of evolution. Nevertheless, over time, surviving parts of the system determine their own self-interest, and that self-interest, kept in balance, yields sustainable coexistence. Likewise, in the case of human function within our ecosystem, our final and ultimate goal must be to leave, as our legacy, a world of land, air, and water resources that is sustainable over time for future generations. To achieve this goal, we must maintain a fundamental balance between rates of utilization and rates of restoration. It is a basic thesis of this report that the equilibration of these two functions can be achieved by management strategies that ensure self-interest through economic incentives and by full participation at the regional level of affected parties, including both industry and the beneficiaries of industry. The concept of restoration must be moved from the cost side to the benefit side of the business ledger. The responsibility for paying the costs of ensuring the benefit of a sustainable environment for future generations must be shared by industry, by government, and by society at large. This goal can be accomplished by applying SEM strategies to carrying out the restoration of environmentally impacted regions.

The concept of sustainable development explicitly recognizes that environmental health, economic vitality, and human welfare are inseparably linked. Hence, sustainable management of ecosystems is closely tied to issues of human population growth, third-world economic development, and environmental justice. These global issues, although overwhelmingly important in the long run, are beyond the scope of this report. The workshop's conclusions and recommendations relate to the more limited issue of using scientific knowledge to effect productive changes in environmental management at local and regional scales.

As scientists, we believe it is clear that solutions to all of the problems discussed in this report are more institutional than scientific.
- The appropriate management scale for most ecosystems is regional rather than local.

- The focus for management decision-making is likely to be regional and local, rather than national.
- Feasible management approaches require involvement of all stakeholders; scientists, engineers, and economists cannot dictate solutions.
- SEM must be an adaptive process in which solutions are tried and modified based on experience.
- The private sector will play a major role in achieving environmental sustainability because 1) the private sector provides many opportunities for environmental improvement through pollution prevention, energy conservation, and improved product design; 2) maintenance of social welfare requires an economically healthy private sector; and 3) large, multinational corporations are ultimately more influential at the global level than are individual governments.
- Enhanced technology transfer is essential, both 1) from federal research and development agencies to state and local regulators and to the private sector and 2) from industrialized nations to developing nations.
- Education concerning the philosophy of and technical approaches to sustainable management is necessary; target audiences include both responsible decision-makers and the public.
- The rate of progress will be enhanced by increased communication between groups involved in the development and application of SEM plans.

For the Society of Environmental Toxicology and Chemistry and the Ecological Society of America, these conclusions have several important implications. Most obvious, research on regional-scale ecological processes is urgently needed, as are geographic information systems and other advanced information technologies for storing, analyzing, and displaying the diverse kinds of data needed to support effective regional management solutions. Both societies are actively promoting this kind of research already. In addition, the involvement of both societies in public affairs is necessary and should be expanded. The ESA Sustainable Biosphere Initiative and the SETAC Congressional Fellowships are important mechanisms for introducing scientific expertise into public decision-making. However, much more emphasis should be placed on interactions with state and local governments, regional cooperatives, and private companies. It is clear that, at least within the United States, these sectors are now the principal focus for innovative management approaches and provide the best opportunities for progress toward environmental sustainability.

We recommend that the two societies should become actively involved in environmental decision-making in a wide variety of contexts. The majority of problems transcend individual disciplines and require multiple talents to devise successful solutions. The effort should transcend traditional employment groups and should involve ecologists and other environmental scientists employed by private companies and nongovernmental organizations and states as well as academic departments and federal agencies. The joint program should include these aspects:

- Active outreach to states and communities and to businesses, offering scientific expertise as needed. In most cases, once an initial network is in place, local scientists can provide most of the needed advice. This implies development of information resources about society members' activities and interests and also conduct of some case studies to demonstrate the value of participatory decision-making involving scientists.
- Educational programs to provide basic information to communities and state leaders and business leaders. Once basic materials are developed, local ESA and SETAC members could distribute them and would be expected to have the best rapport with local leaders.
- Promote involvement of scientists in developing countries. More active technology transfer is needed but requires the support of the U.S. government or major private organizations and is beyond the resources or organizational capabilities of ESA and SETAC.

Above all, scientists within both societies must understand that, as implied by the management "circle" in Figure 2 (Chapter 1), sustainable environmental management is not a linear process in which experts (e.g., ecologists and environmental toxicologists) prescribe solutions that are then enacted by governments or corporate executives. Rather, sustainable environmental management is a cyclical process in which scientists play a complex role that involves education as well as technical analysis, and in which decisions are ultimately made by the people directly affected by them.

Appendix A

Workshop Agenda

Wednesday 8/25/93

(evening)
Workshop objectives, announcements
Larry Barnthouse, Rod Parrish

Thursday 8/26/93

(morning)
Issue papers and commentaries:

The Sustainable Biosphere Initiative
Marge Holland, Ecological Society of America
Richard Haeuber, Sustainable Biosphere Initiative

Risk Assessment and Sustainable Environmental Management
Greg Biddinger, Exxon Company USA

Institutional Constraints on Sustainable Environmental Management
William Mulligan, Chevron Corporation

Economic Valuation of Ecological Resources
Paul Portney, Resources for the Future

(afternoon)
Initial meetings of breakout groups:

Energy/transboundary pollution
(*Terry Yosie*, chair)

Integrated landscape/watershed management
(*Marge Holland*, chair)

Product stewardship
(*Jim Fava*, chair)

Preservation of biodiversity
(*Bill Cooper*, chair)

Urban environmental management
(*Jim Gillett*, chair)

Environmental restoration
(*Greg Biddinger*, chair)

Friday 8/27/93

(morning)

Initial set of case studies, representing best current state-of-the-art in sustainability-based environmental management (40-minute presentation, commentaries, 30-minute discussion for each)

Integrated pest management
Joseph Kovach, Cornell University

Forest Management
Dean Premo, White Water Associates

Everglades
Larry Harris, University of Florida

(afternoon)

Breakout group sessions

Saturday 8/28/93

(morning)

Second set of case studies: new problems in sustainable environmental management

Great Lakes
John Hartig, Wayne State University

Energy Production Systems
Panel discussion,
chaired by *Terry Yosie, E. Bruce Harrison Co.*

Product Stewardship
Sven-Olof Ryding, Federation of Swedish Industries

(afternoon)

Breakout group sessions

(evening)

Plenary session: progress reports from working groups

Sunday 8/29/93   **(all day)**
Breakout group sessions, report writing

Monday 8/30/93   **(morning)**
Plenary session: presentation and discussion of working group reports

(afternoon)
Steering Committee drafts report

Tuesday 8/31/93   **(morning)**
Steering Committee drafts report

Appendix B

Workshop Participants and Contributing Authors

Larry Barnthouse
Oak Ridge National Laboratory
Oak Ridge TN

Steven M. Bartell
SENES Oak Ridge, Inc.
Center for Risk Analysis
Oak Ridge TN

Gregory R. Biddinger
Exxon Biomedical Sciences, Inc.
East Millstone NJ

Allegra Cangelosi
Great Lakes Washington Program
Northeast Midwest Institute
Washington DC

Frank J. Consoli
Scott Paper Company
Philadelphia PA

William Cooper
Institute for Environmental Toxicology
Michigan State University
East Lansing MI

Peter deFur
Environmental Defense Fund
Washington DC

Elaine J. Dorward-King
Kennecott Corporation
Salt Lake City UT

Chuck DuMars
University of New Mexico School of Law
Albuquerque NM

James A Fava
Roy F. Weston Inc.
West Chester PA

Steven M. Fetter
Public Service Commission
Lansing MI

James Gillett
Department of Natural Resources
Cornell University
Ithaca NY

Rick Haeuber
Sustainable Biosphere Initiative
Washington DC

Larry D. Harris
University of Florida, School of
    Forest Resources and Conservation
Gainesville FL

John Hartig
Wayne State University
Chemical Engineering
Detroit MI

Marge Holland
Ecological Society of America
Washington DC

Robert J. Huggett
Virginia Institute of Marine Science
College of William and Mary
Gloucester Point VA

Joseph Kovach
NYS Agr Exp Station
Integrated Pest Management
Geneva NY

Michael E. Kraft
Public and Environmental Affairs
University of Wisconsin-Green Bay
Green Bay WI

SETAC Press

William Mulligan
Chevron Corporation
Washington DC

Rodney Parrish
SETAC/SETAC Foundation
Pensacola FL

Paul R. Portney
Resources for the Future
Washington DC

Dean Premo
White Water Associates
Amasa MI

Douglas P. Reagan
Woodward-Clyde Consultants
Castle Rock CO

Kevin Reinert
Department of Toxicology
Rohm & Haas Company
Spring House PA

Geraldo Stachetti Rodrigues
CNPMA/EMBRAPA (BRAZIL)
Boyce Thompson Institute
 for Plant Research
Cornell University
Ithaca NY

Sven-Olof Ryding
Federation of Swedish Industries
Stockholm, Sweden

Glenn W. Suter
Environmental Sciences Division
Oak Ridge National Laboratory
Oak Ridge TN

Frieda B. Taub
School of Fisheries
University of Washington
Seattle WA

Judith S. Weis
Department of Biological Sciences
Rutgers University
Newark NJ

Tse-Sung Wu
Carnegie Mellon University
Dept. of Engineering and Public Policy
Pittsburgh PA

Terry F. Yosie
E. Bruce Harrison Company
Washington DC

References

Allenby BR. 1994. Integrating environment and technology: Design for Environment. In: Allenby BR, Richard D, editors. 1994 greening industrial ecosystems. Washington DC: National Academy Pr.

Allenby BR, Fullerton A. 1991–92. Design for the Environment: A new strategy for environmental management. *Pollution Prevention Review*: Winter.

Canada-Ontario 1988. Remedial action plan for Hamilton Harbour: goals, problems, and options. Burlington, Ontario. 200 p.

Crombie D. 1990. Waterfront: interim report. Toronto, Ontario: Royal Commission on the Future of the Toronto Waterfront. 207 p.

Endangered Species Act. Title 16 § 1531. December 28, 1973.

Fava JA, Adams WJ, Larson RJ, Dickson GW, Dickson KL, Bishop WE, editors. 1987. Research priorities in environmental risk assessment. Pensacola FL: SETAC Pr. 103 p.

Fava JA, Dennison R, Jones B, Curran MA, Vigon B, Selke S, Barnum J. 1991. A technical framework for life-cycle assessment. SETAC Pellston Workshop; 1990 Aug 18–23; Smuggler's Notch VT. Pensacola FL: SETAC Pr.

Fava JA, Consoli F, Dennison R, Dickson K, Mohin T, Vigon B. 1993. A conceptual framework for life-cycle impact assessment. SETAC Pellston Workshop; 1992 Feb 1–6; Sandestin FL. Pensacola FL: SETAC Pr.

Giesy JP, Ludwig JP, Tillitt DE. 1994. Deformities in birds of the Great Lakes Region. *Environ Sci Technol* 28:128A-135A.

Gillett JW, Thompson JB, Urey DG, Costanza R, Tinsley IJ, Weis JS, Yanders AF. 1992. The need for an integrated urban environmental policy. *J Urban Affairs* 14:377–398.

Great Lakes Fishery Commission. 1980. A joint strategic plan for management of Great Lakes fisheries. Ann Arbor MI. 23 p.

Hartig JH, Fuller K, Epstein D, Coape-Arnold T, Hottman A. 1993. Great Lakes RAPs are a hit. Water Environment and Technology. p 52–57.

Hartig JH, Law NL. 1994. Institutional frameworks to direct the development and implementation of remedial action plans. *Environ Management* 18(6):855–864.

Hartig JH, Zarull MA. 1992. Under RAPs: toward grassroots ecological democracy in the Great Lakes Basin. Ann Arbor MI: Univ of Michigan Pr. 289 p.

Healthy Toronto 2000 Subcommittee. 1988. Healthy Toronto 2000: a strategy for a health city. Toronto, Ontario: Dept of Public Health. 19 p.

Holland MM. 1993. Management of land/inland water ecotones: needs for regional approaches to achieve sustainable ecological systems. *Hydrobiologia* 251: 331–340.

Holland MM. 1996. Sustainability of natural resources: focus on institutional mechanisms. *Can J Fish Aquat Sci* 53:432–434.

Holling CS, editor. 1978. Adaptive environmental assessment and management. Chichester UK: J Wiley.

Hotchkiss BE, Gillett JW, Kamrin MA, Witt JW, Craigmil A. 1989. EXTOXNET, Extension Toxicology Network. A pesticide information project of cooperative extension offices of Cornell University, the University of California, Michigan State University, and Oregon State University. Ithaca NY: Cornell Univ.

Houghton JJ, Meiro Filho LG, Callander BA, Harris N, Kattenberg A, Maskell K. 1995. Climate change 1995 - the science of climate change. Contribution of Working Group I to the Second Assessment Report of the Intergovernmental Panel on Climate Change. Cambridge UK: Cambridge Univ Pr.

Huggett RJ, Kimerle RA, Mehrle PM, Bergman HL, editors. 1992. Biomarkers: biochemical, physiological, and histological markers of anthropogenic stress. Chelsea MI: Lewis. 347 p.

Jaworski E, Raphael CN. 1978. Fish, wildlife, and recreational values of Michigan's coastal wetlands (Phase I): Coastal wetlands value study in Michigan. Lansing MI: Michigan Department of Natural Resources (MDNR). 209 p.

Kendall RJ and Lacher Jr TE, editors. 1994. Wildlife toxicology and population modeling: integrated studies of agroecosystems. Boca Raton FL: Lewis. 576 p.

Lawrence DM, Holland MM, Morrin DJ. 1993. Profiles of ecologists: results of a survey of the membership of the Ecological Society of America. Part I. A snapshot of survey respondents. *Bull Ecol Soc Am* 74:21–35.

Lubchenco J, Olson AM, Brubaker LB, Carpenter SR, Holland MM, Hubbell SP, Levin SA, MacMahon JA, Matson PA, Melillo JM, Mooney HA, Peterson CH, Pulliam HR, Real LA, Regal PJ, and Risser PG. 1991. The sustainable biosphere initiative: an ecological research agenda. *Ecology* 72(2): 371–412.

Ludwig DL, Hilborn R, Walters C. 1993. Uncertainty, resource exploitation, and conservation: lessons from history. *Science* 260:17, 36.

Kovach J, Petzolt C, Degni J, Tette J. 1992. A method to measure the environmental impact of pesticides. *NY Food Life Sci Bull* 139:1–8.

Mills EL, Leach JH, Secor CL, and Carlton JT. 1993. What's next? the prediction and management of exotic species in the Great Lakes (report of the 1991 workshop). Great Lakes Fisheries Commission. 22 p.

Mladenoff DJ, Pastor J. 1993. Sustainable forest ecosystems in the northern hardwood and conifer region: concepts and management. In: Aplet GH, Olson JT, Johnson N, Sample VA, editors. Defining sustainable forestry. Washington DC: Island Pr.

National Research Council. 1983. Risk assessment in the federal government: managing the process. Washington DC: National Academy Pr.

National Research Council. 1993. Issues in risk assessment. Volume 3: Ecological risk assessment. Washington DC: National Academy Pr.

[NEPA] National Environmental Policy Act. Title 42 § 4321. January 1, 1970.

Noss RF, Harris LD. 1986. Nodes, networks, and MUMs: preserving diversity at all scales. *Environ Management* 10:299–309.

Pacific Rivers Council. 1993. Entering the watershed: an action plan to protect and restore America's river ecosystems and biodiversity. Washington DC: Island Pr.

Premo D. 1992. A prospectus for the Mulligan Creek Riparian Management Area. Amasa MI: White Water Associates. 9 p.

Rogers E, Premo D. 1992. Total ecosystem management strategies: a study guide. Amasa MI: White Water Associates, Inc. 45 p.

Ryding S-O. 1992. Environmental management handbook. Amsterdam, The Netherlands: IOS Pr. 777 p.

Schmidheiny S. 1992. Changing course: a global business perspective on development and the environment. Cambridge MA: MIT Pr. 374 p.

Slocombe DS. 1993. Environmental panning, ecosystem science, and ecosystem approaches for integrating environment and development. Environ Manage 17:289–303.

Smith WG, Barnard J. 1992. Chem-New Profiles. Pesticide Management and Education Program, CNET, Cornell Cooperative Extension Electronic Information Network. Ithaca NY: Cornell Univ.

[SMCRA] Surface Mining Control and Reclamation Act. 1977. U.S. Public Law 95-87.

Starfield AM, Bleloch AL. 1991. Building models for conservation and wildlife management. Edina MN: Burgess International Group. 253 p.

Steen B, Ryding S-O. 1992. The EPS Enviro-Accounting method: an application of environmental accounting principles for evaluation and valuation of environmental impact in product design.

Göteborg, Sweden: Swedish Environmental Research Institute, Report B 1080. Available from: Swedish Environmental Research Institute, Box 47086, S-40258, Göteborg, Sweden.

Strachan WM, Eisenreich SJ. 1988. Mass balancing of toxic chemicals in the Great Lakes: the role of atmospheric deposition. Windsor, Ontario: Report of the International Joint Commission's Water Quality Board.

Theiling KM, Croft BA. 1988. Pesticide side-effects on arthropod natural enemies: a database summary. *Agric Ecosyst Environ* 21:191–218.

[USEPA] U.S. Environmental Protection Agency. 1992. Framework for ecological risk assessment. Washington DC: USEPA. EPA/630/R-92/001.

Walters C. 1986. Adaptive management of renewable resources. New York: MacMillan. 374 p.

Waybrant JR, Bryant WC. 1990. A management plan for the sport fishery of the St. Clair-Detroit Rivers system and Lake Erie. Lansing MI: Michigan Department of Natural Resources (MDNR), Fisheries Division. 21 p.

Westman WE. 1985. Ecology, impact assessment and environmental planning. New York: J Wiley. 532 p.

Whitaker JB. 1993. Launching the Great Lakes Initiative. *Water Environ Technol* 5(6):40–46.

White Water Associates Inc. 1992a. Riparian management area. Strategies: an update bulletin for the TEMS program 1(1). Amasa MI: White Water Associates.

White Water Associates Inc. 1992b. The woodland vernal pond: an oasis of diversity. Strategies: an update bulletin for the TEMS program 1(3). Amasa MI: White Water Associates.

White Water Associates Inc. 1993. Mulligan Creek focus of new public-private resource coalition. Strategies: an update bulletin for the TEMS program 2(1). Amasa MI: White Water Associates.

World Commission on Environment and Development. 1987. Our common future. Oxford UK: Oxford Univ Pr. 400 p.

Index

A

abbreviations and acronyms xxiii–xxiv
academic institutions 12
accounting
    Enviro-Accounting Method 36
    full cost 51
acid deposition
    landscape level of 9
    transboundary issues xx, 8
adaptive management 6, 72–76
agriculture 8, 14
    apple production 21–22
    deforestation for 22
    Florida Everglades 39
    integrated pest management xx, 14, 19–22
    nonpoint source pollution from 9, 13
    sustainable, examples of 19
    traditional *vs.* IPM and organic pest management 21–22
air quality. *See also* transboundary issues
    emissions 34, 35, 64
    purification by ecosystems 14
    transboundary issues and xix, 7–9, 42
alien species. *See* nonindigenous species
aluminum beverage containers 48
aquatic habitats 13, 61
avian population ecology viii

B

Best Management Practices 61
bioaccumulation 24
biodiversity "back-feeds" 14
biodiversity protection and restoration xi, xx, xxi, 13–16. *See also* nonindigenous species
    Florida Everglades 37
    Michigan, Upper Peninsula 29–32
    reserves 13, 59, 60
    social choice mechanisms and 57–62
    two schools of strategies for 15
biomarkers viii
birds
    Florida Everglades 37, 38
    Michigan, Upper Peninsula 30–31
buffer zones in forest management 30, 31

C

Canada
    acid deposition and transboundary issues 8
    Ontario Roundtables 26
    phosphorous control program 24
    Virtual Elimination Task Force 25
case studies
    Florida Everglades xx, 4, 37–40
    Great Lakes xx, xxi, 4, 22–28, 64
    integrated pest management xx, 14, 19–22
    Michigan forest management 28–33
    Sweden, product development xx, 33–36
CFC (chlorofluorocarbons) 12
Chalmers University of Technology 34
champion, appointment of a 52, 53
Champion International 32
channelization 9, 10, 39
chemical manufacturers 12
Chemical Manufacturers Association (CMA) 12, 49
CHEM-NEWS 20
chlorofluorocarbons (CFC) 12
cities. *See* urban environmental management; urban runoff
CNET (Cornell Cooperative Extension Network) 20
coastal habitat
    of the Great Lakes 23, 24
    ownership issues and public domain 61
command-and-control approach 5, 17, 46
    for environmental restoration xxi, 70–72
    as a regulatory strategy 4, 79
commitment 46, 59
communication 54, 55, 78, 80
    in an Ecosystem Management Strategy Agreement 44–45
    between ecologists and foresters 29, 30–33
    in product development 36
community, sense of 66. *See also* public participation
complexity, of management actions 4
concurrent design 49
connectivity of ecosystems 10
construction sites 9
conventional strategies, *vs.* sustainable environmental management 4, 17, 68–70

coordination
  cross-disciplinary teams 49
  of multi-stakeholders 27, 64
coral reefs 13
Cornell Cooperative Extension Network (CNET) 20
Cornell University IPM Program 20
corridors. *See* migration routes
cross-disciplinary teams 49
cultural aspects 3, 16–17

D

dams 9, 10
data transfer 9
debt-for-nature 62
decision-making process xi, xx, 13, 53–54, 68–69. *See also* sociopolitical aspects
  in forest management 29
  inclusiveness in the 4, 12, 26, 27
  organizational aspects of 54
  regional and local 80
  remedial action plan in the 26
  role of science in the 2, 5–6, 26
deforestation 19, 22, 23
demographics 3, 64
Design for Sustainable Development (DFSD) 47–48
Design for the Environment (DFE) 12, 49
developing countries. *See* less developed countries
development, habitat loss from 23
DFE (Design for the Environment) 12, 49
DFSD (Design for Sustainable Development) 47–48
dispersal. *See* migration routes
disturbances, ecological role of 15
DOE (U.S. Department of Energy) 78
drainage basins 9, 38
dredging 23

E

EA (environmental audit) 50
Eastern Europe, transboundary issues 8
ecodiversity 14
ecolabeling 50
Ecological Society of America (ESA) ix–x, xix, 43
  Public Affairs Office x
  workshops viii, xi–xii. *See also* Pellston Workshop on Sustainability-Based Environmental Management
economic aspects 2, 3, 11, 41, 42, 54
  budget cycles and environmental planning 46
  costs of environmental restoration 67–68, 69–70
  Enviro-Accounting Method 36
  full cost accounting 51
  innovations in reconfiguring ownership 62
  manufacturing process and 2, 11
  social choices and 57–59
  tax incentives 60, 70, 71, 72, 76–77
  "willingness to pay" concept 35
Ecosystem Management Strategy Agreements (EMSA) 44–47
ecosystems
  disturbances in 15
  six properties of functional 9–10
  "snapshot of the past" approach, limitations of 69
  water and air purification by 14
ECU (European currency unit) 35
education 5, 27, 54, 80, 81
  business/organizational 13, 54, 55, 66
  cooperative field trips 29
  on forestry 29
  on life cycle assessment in product development 36
  on preventive *vs.* curative tasks 46
  role in implementation 56, 78
  of stakeholders xxii, 13, 53
  on sustainable development 41
  in urban environmental management 57
EIA (environmental impact assessment) 34, 36, 49–50
EIQ (environmental impact quotient) 20–22
emissions. *See also* transboundary issues
  assessment of, in product development 34, 35
  in urban environments 64
EMSA (Ecosystem Management Strategy Agreements) 44–47
Endangered Species Act of 1973 60, 72
energy
  agriculture and intensive use of 19
  conservation xxii

energy (*continued*)
    extraction and refining 8
    flowchart for product life cycle 48
    in life-cycle assessment 35, 42, 48, 51
    in urban environments 63
energy efficiency 8
engineering 55
Enviro-Accounting Method 36
environmental audit (EA) 50
environmental impact assessment (EIA) 34, 36, 49–50
environmental impact quotient (EIQ) 20–22
environmental labeling 50
environmental load index 34, 35–36
environmental management
    proactive *vs.* reactive 22
    protection 2, 11
    vs. environmental development 2
Environmental Priority Strategies (EPS) 34–38
environmental restoration xx, 18, 67–79
    adaptive management and 74
    command-and-control approaches to xxi, 46, 70–72
    Florida Everglades 39–40
    prioritization 77
EPA. *See* U.S. Environmental Protection Agency
erosion 9, 23, 24
error analysis 34, 36
estuaries 13, 37
European currency unit (ECU) 35
eutrophication 22, 24
Everglades. *See* Florida Everglades, case study
Everglades National Park 39
executive summary xix–xxii
exotic species. *See* nonindigenous species
exposure assessment 51. *See also* risk assessment
exposure-response assessment 51
Extension Toxicology Network (EXTOXNET) 20

F

family aspects 3
Federation of Swedish Industries 34
figures, list of vii
fish viii
    Florida Everglades 37, 38
    Great Lakes 23, 25–26

flooding
    Florida Everglades 38
    Great Lakes region 24
    transboundary issues and xix, 7, 8
Florida Everglades, case study xx, 4, 37–40
forest management 15
    complexity of management actions and 4
    Michigan case study 28–33
    vernal ponds and 31–32
fragmentation of habitat 10, 15
full cost accounting 51

G

game management 15
"gate-to-gate" perspective 11
genetic aspects 10, 14
geographical information system (GIS) 29, 33, 61, 79
global climate change viii, xi, 25, 38, 39
global-level issues 2, 18
governmental agencies 12
    development of commitment by 15
    goals of, in regional management 43–44, 45
    public policy limitations of 57–59
    technology transfer and 43
Great Lakes, case study xx, xxi, 4, 22–28, 64
Great Lakes Fishery Commission 25
Great Lakes Initiative 26–27
Great Lakes Toxic Substances Control Agreement 46
Great Lakes Water Quality Agreement 25, 26, 44, 46
"greenhouse effect." *See* global climate change

H

habitat destruction viii, 9
    Florida Everglades 38–39
    Great Lakes region 23
habitat diversification 14
habitat fragmentation 10, 15
hazard identification 52. *See also* risk assessment
hazardous wastes
    barges of, and docking privileges 8
    transboundary issues xix, 8
health care 66
historical background viii–ix, xix
    Florida Everglades 37
    Great Lakes region 22, 23

historical background (*continued*)
    need for information base 8
    urban environmental management 17
"hot spots" 15, 61, 62
human resources 11, 54, 55
hydrology, modified 9, 10, 37–39

I

implementation 52, 53–54, 56–57, 70–72
incentives, economic 60, 66, 70, 71, 72, 76–77
inclusiveness of management decisions 4, 12, 26, 27
India, transboundary issues and flooding 8
industry. *See* private sector
information transfer 77–79
    between ecologists and foresters 29–33
    in education programs 29–33
    to stakeholders 53
integrated pest management (IPM) xx, 14, 19–22
interdisciplinary cooperation xi, xix
intergovernmental cooperation 44–47
international factors 25, 41, 42, 43. *See also* transboundary issues
International Joint Commission 25
introduced species. *See* nonindigenous species
inventory 34, 50
IPM (integrated pest management) xx, 14, 19–22

J

jurisdictional issues xix, 8, 41, 60–62

L

land ownership 15, 25, 58, 60–62
landscape level xix, 9–10, 29–30, 32–33
LCA. *See* life-cycle assessment
LDC. *See* less developed countries
legal aspects 3, 11. *See also* sociopolitical aspects
    of biodiversity protection xxi
    military land 60
    ownership rights 15, 26, 58, 60–62

less developed countries (LDC). *See also* technology transfer
    information transfer 77
    issues of, *vs.* developed nations 2, 18
    technology assistance for 42, 81
    transboundary issues of 8
levees 9, 10
Life-Cycle Assessment Advisory Group 4
life-cycle assessment (LCA) 3, 12, 51
    energy flowchart 48
    in product development case study 33–36
logging 13, 19
Longyear Realty 32
lynx, Canadian 32

M

mammals
    Florida Everglades 39
    forest management and 31, 32
manufacturing process. *See* process design
marinas 9
marketing/sales 11
    biodiversity and 16, 57–62
    environmental labeling 50
    identification of opportunities 42
    product stewardship and 47
Massachusetts, Water Supply Citizens Advisory Committee (WSCAC) 44
materials acquisition 11, 48, 51, 54, 55
Mead Corporation, forest management case study 28–33
mercury contamination 38, 39
Mexico, transboundary issues 8
Michigan. *See also* Great Lakes, case study
    bird diversity 30–31
    forest management case study 28–33
    mammals 31, 32
    percent of Great Lakes coastal wetland lost 23
Michigan Department of Natural Resources 33
Michigan State University 20
migration routes 9, 15
    Florida Everglades 39
    Michigan, Upper Peninsula 31, 32
mining and minerals extraction 8
mission statement 53, 54
mitigation land banks 62
modeling 10, 78. *See also* risk assessment; uncertainty

monitoring
    of pests 22
    of sustainable environmental management 6, 27, 45–46, 56, 66
Mulligan Creek Coalition 32–33
mussel, zebra 23

N

NAFTA (North American Free Trade Agreement) 8
National Aeronautic and Space Administration (NASA) x
National Environmental Policy Act of 1970 (NEPA) 18
National Oceanic and Atmospheric Association (NOAA) 78
national parks 59, 62
National Research Council 51
National Science Foundation x–xi
native species, in maintenance of gene pool 10
Nature Conservancy 58, 61
Nepal, transboundary issues and flooding 8
NEPA (National Environmental Policy Act of 1970) 18
NGO (nongovernmental organizations) 12, 44
nitrogen fixation 14
NOAA (National Oceanic and Atmospheric Association) 78
nongovernmental organizations (NGO) 12, 44
nonindigenous species 9, 16
    in the Florida Everglades 39
    in the Great Lakes region 23
    integrated pest management and 19
nonpoint source pollution 9, 13
North American Free Trade Agreement (NAFTA) 8

O

Oregon State University 20
organizational aspects 52–56, 64–66
outreach programs 27, 46, 83
ownership rights 15, 25, 58, 60–62

P

panther, Florida 39
paper manufacture and sustainable forest management xx

PCBs (polychlorinated biphenyls) 24
Pellston Workshop on Sustainability-Based Environmental Management
    agenda 85–87
    conclusions and recommendations xxi–xxii, 81–83
    foundation for cooperation from xi
    major themes xix–xx
    participants and contributing authors 89–90
    recommendations of breakout groups xx–xxi
    steering committee xviii
pesticides
    contamination, viii 19, 20
    environmental impact quotients 20–21
    role in integrated pest management 20
    USEPA Fact Sheets 20
phosphorous pollution
    in the Florida Everglades 39
    U.S.-Canada control program 24
plants, of the Florida Everglades 37
policy formation 56–57, 58–60. *See also* decision-making process
political aspects. *See* sociopolitical aspects
pollution 13, 14
    emissions 34, 35, 64
    mercury 38, 39
    from nonpoint sources 9, 13
    pesticides viii, 19, 20
    phosphorous 24, 39
    polychlorinated biphenyls 24
    transboundary issues viii, xix, 7–9, 42
pollution prevention xxii, 50, 66
polychlorinated biphenyls (PCBs) 24
population density 41, 64
pre-disturbance planning 73
prioritization 27, 66
    in adaptive management 73, 74
    of ecosystems to address 1–2
    in environmental restoration 77
    of opportunities for improvement 55
private sector. *See also* product stewardship; technology development
    development of commitment 15
    "gate-to-gate" perspective of 11
    inclusiveness with the public 4, 70
    role of xxii, 44, 80
problem-solving process 1
process design 2, 11, 35, 40, 54
procurement 11, 48, 51, 54, 55

SETAC Press

product stewardship xix, xxi, 4, 11–13, 47–52.
    See also Sweden, product development
        case study
property rights 15, 25, 58, 60–62
psychosocial aspects 3
public domain 61–62
public outreach 27, 46, 83
public participation 3, 5, 17, 44–45, 69–70
    in the Great Lakes Initiative 27
    in implementation 53, 54, 66
    in urban environmental sustainability xxi
public perception/awareness 5, 23. *See also*
        education
    development of trust 59–60
    environmental labeling and 50
public-private partnerships 59, 66, 70
public relations 11
public-trust doctrines 58, 60

R

rain forests 13
RAPs (remedial action plans) 26, 64, 64, 65
raw materials, in life cycle assessment 34, 35
reclamation 67
recreation value 23, 39
recycling/re-use 34, 48, 49, 56
references 91–93
regional and local approach 68–69, 72–73
regulations. *See also* sociopolitical aspects
    adaptive management and 6, 72–76
    approaches for urban environmental
        management 64–66
    command-and-control approach to 4, 79
    compliance 11
    Endangered Species Act of 1973 60, 72
    government affairs 11
    Great Lakes Initiative 26–27
    National Environmental Policy Act of 1970 18
    overcoming inertia 79
    permitting, integration of 64
    regional or local solutions 68–69, 72–73
    Resource Conservation and Recovery Act of
        1976 69
    Surface Mining Control and Reclamation Act
        of 1977 18
    Swamp and Overflowed Lands Law of 1850
        37

remedial action plans (RAPs)
    Great Lakes region 26
    sustainable urban environments and 64
remediation 42, 67
    cost of 51
    vs. prevention 46
reptiles, of the Florida Everglades 37
research 66
    cooperative, between ecologists and foresters
        30–33
    and development 11, 54, 55
    integration of 27, 55
    international aspects 43
    scale of investigation and 15
reserves 13, 59, 60
resilience of ecosystems 10
Resource Conservation and Recovery Act of 1976
    69
resource extraction. *See* life-cycle assessment;
        product stewardship
resources
    agriculture and intensive use of 19
    in full cost accounting 51
    identification of 5
    use of, in life cycle assessment 34, 35
resource stewardship 58
Responsible Care program 12, 49
restoration. *See* biodiversity protection and
        restoration; environmental restoration
riparian management 13
    dams 9, 10
    Michigan forest management case study 31,
        32–33
    stream diversions 9, 10
risk assessment 22, 51–52
    need for predictive capability 10
    for prioritization 27
    sensitivity and error analysis 34, 36
    uncertainty factors 51
risk characterization 51. *See also* risk assessment
rivers. *See* riparian management

S

scale, of management actions 4, 6, 27
    biodiversity and 15
    budget cycles and, 46
    Florida Everglades 37–38

science, role of
  in creation of an information base 8, 78
  in environmental management development 1–2, 5–6, 26, 45, 46, 83
  in forest management 28
  in urban environmental management 62–64
sea level rise 38, 39
sectors, of management 64, 66
sediment contamination 9, 23
SELCTV database 20
self-sustainability of ecosystems 10
SETAC. *See* Society of Environmental Toxicology and Chemistry
sewage treatment 9
SIMPLE (Sustainability of Intensively Managed Populations in Lake Ecosystems) 25–26
site-specific environmental management 22, 64
SMCRA (Surface Mining Control and Reclamation Act of 1977) 18
social aspects
  choice mechanisms and biodiversity 57–62
  private sector and 82
social responsibility 2, 9
  impact categories 3
  in the manufacturing process 11
Society of Environmental Toxicology and Chemistry (SETAC) ix, xix, 43
  Congressional Fellowships 82
  Life-Cycle Assessment Advisory Group 4
  workshops viii, xi, xviii, 3. *See also* Pellston Workshop on Sustainability-Based Environmental Management
sociocultural aspects 3, 16–17
sociopolitical aspects 3, 28, 46, 59
  budget cycles and environmental planning 46
  continuity through electoral transitions 46
  Florida Everglades 39
  identification of resources and services 5
  inclusiveness and 4
  of urban environmental management 17
soil ecology 14
spatial extent of management. *See* scale, of management actions
stakeholders 82
  in adaptive management 73, 74, 75
  assessment of 54
  communication and education of 53
  multi- 27, 64
stream diversions 9, 10

success, celebration of 27
Surface Mining Control and Reclamation Act of 1977 (SMCRA) 18
Sustainability of Intensively Managed Populations in Lake Ecosystems (SIMPLE) 25–26
Sustainable Biosphere Initiative ix, x–xi, 1, 82
sustainable development
  definition of 1, 53
  strategic matrix for 55
"Sustainable Development Corps" 43
sustainable environmental management. *See also* individual case studies
  adaptive nature of 6, 72–76
  definition 2
  global, and dissemination of sustainable technologies 41–43
  implementation 52, 53–54, 56–57, 70–72
  of regions, strategy for 43–47
  six properties of functional ecosystems 9–10
  strategy development and 5–6, 12–13
  ten steps to 52–56
  three principal components 2–3
  vs. conventional strategies 4, 17, 67–70
Sven-Olof Ryding, product development case study xx, 33–36
Swamp and Overflowed Lands Law of 1850 37
Sweden, product development case study xx, 33–36
Swedish Environmental Research Institute 34

T

tables, list of vii
technology development
  identification of markets 42
  product design xxii
  product stewardship and xix, xxi
  research and development for 11, 54, 55
  role of science in 1–2, 5–6
technology transfer 41–43
  among stakeholders xxi, 77–78, 82
  to less developed nations xx–xxi, xxii, 8, 82
  mechanisms for 77–79
TEMS (Total Ecosystem Management Strategies) 29
timber harvest 13, 19
timeframe. *See* commitment; scale, of management actions
Total Ecosystem Management Strategies (TEMS) 29

Total Quality Management (TQM) 13, 49, 56, 73
toxic substances
   Florida Everglades 38, 39
   Great Lakes region 24, 25, 26–27
   transboundary issues 8
transboundary issues viii, xix, 7–9, 42
transportation, urban 16–17

U

uncertainty
   in environmental management 10, 27, 28, 66
   in life-cycle assessment for product development 34, 36
   in risk assessment 51
United Nations Conference on the Environment and Development (UNCED) 1
unit effects 35
University of California 20
Upper Peninsula. *See* Michigan, forest management case study
urban environmental management xx, xxi, 16–17, 62–66
urban runoff 9
U.S. Department of Energy (USDOE) 78
U.S. Environmental Protection Agency (USEPA) 78
   Environmental Monitoring and Assessment Program 61, 79
   Pesticide Fact Sheet 20
   pesticide registration 20
U.S. Fish and Wildlife Service (FWS) 61, 78
U.S. Forest Service x
U.S. Geological Survey (USGS) 61, 78
U.S. National Park Service 59, 62
usuary rights 60

V

vernal ponds 31–32
Virtual Elimination Task Force 25

W

waste disposal
   cost of 51
   sustainable 69

water level fluctuations
   Florida Everglades 37–39
   Great Lakes region 24
   in sea level, and global climate change 38, 39
water quality. *See also* transboundary issues
   Great Lakes Water Quality Agreement 25, 26
   Michigan forest management case study 30
   purification by ecosystems 14
watershed management xix, 9–10, 43–44, 46
   Florida Everglades 38
   Great Lakes region 27
water supplies
   Florida Everglades 39
   hydroperiods and 10, 39
   transboundary issues xix
Water Supply Citizens Advisory Committee (WSCAC), 44
wetlands
   Florida Everglades xx, 4, 37–40
   Great Lakes region 23
White Water Associates, Inc., forest management case study 28–33
WHO (World Health Organization) 16
wilderness areas 59
wildlife corridors. *See* migration routes
wolf, grey 32
World Commission on Environment and Development 1
World Health Organization (WHO) 16

Z

zoning 61

SETAC

A Professional Society for Environmental Scientists and Engineers and Related Disciplines Concerned with Environmental Quality

The Society of Environmental Toxicology and Chemistry (SETAC), with offices in North America and Europe, is a nonprofit, professional society that provides a forum for individuals and institutions engaged in the study of environmental problems, management and regulation of natural resources, education, research and development, and manufacturing and distribution.

Goals
- Promote research, education, and training in the environmental sciences
- Promote systematic application of all relevant scientific disciplines to the evaluation of chemical hazards
- Participate in scientific interpretation of issues concerned with hazard assessment and risk analysis
- Support development of ecologically acceptable practices and principles
- Provide a forum for communication among professionals in government, business, academia, and other segments of society involved in the use, protection, and management of our environment

Activities
- Annual meetings with study and workshop sessions, platform and poster papers, and achievement and merit awards
- Monthly scientific journal *Environmental Toxicology and Chemistry*, SETAC newsletter, and special technical publications
- Funds for education and training through the SETAC Scholarship/Fellowship Program
- Chapter forums for the presentation of scientific data and for the interchange and study of information about local concerns
- Advice and counsel to technical and nontechnical persons through a number of standing and ad hoc committees

Membership

SETAC's growing membership includes more than 5000 individuals from government, academia, business, and public-interest groups with technical backgrounds in chemistry, toxicology, biology, ecology, atmospheric sciences, health sciences, earth sciences, and engineering.

If you have training in these or related disciplines and are engaged in the study, use, or management of environmental resources, SETAC can fulfill your professional affiliation needs. Membership categories include Associate, Student, Senior Active, and Emeritus.

For more information, contact SETAC, 1010 North 12th Avenue, Pensacola, Florida, USA 32501-3370; T 850 469 1500; F 850 469 9778; E setac@setac.org; http://www.setac.org.